The Metaphysics of Contingency

Mind, Meaning and Metaphysics

Series Editors:

Johannes L. Brandl, University of Salzburg, Austria
Christopher Gauker, University of Salzburg, Austria
Max Kölbel, University of Vienna, Austria
Mark Textor, King's College London, UK

The Mind, Meaning and Metaphysics series publishes cutting-edge research in philosophy of mind, philosophy of language, metaphysics and epistemology. The basic questions in this area are wide-ranging and complex: What is thinking and how does it manage to represent the world? How does language facilitate interpersonal cooperation and shape our thinking? What are the fundamental building blocks of reality, and how do we come to know what reality is?

These are long-standing philosophical questions but new and exciting answers continue to be invented, in part due to the input of the empirical sciences. Volumes in the series address such questions, with a view to both contemporary debates and the history of philosophy. Each volume reflects the state of the art in theoretical philosophy, but also makes a significant original contribution to it.

Editorial Board:

Annalisa Coliva, University of California, Irvine, USA
Paul Egré, Institut Jean-Nicod, France
Olav Gjelsvik, University of Oslo, Norway
Thomas Grundmann, University of Cologne, Germany
Katherine Hawley, University of St. Andrews, United Kingdom
Øystein Linnebo, University of Oslo, Norway
Teresa Marques, University of Barcelona, Spain
Anna-Sophia Maurin, University of Gothenburg, Sweden
Bence Nanay, University of Antwerp, Belgium
Martine Nida-Rümelin, University of Freiburg, Switzerland
Jaroslav Peregrin, Czech Academy of Sciences, Czech Republic
Tobias Rosefeldt, Humboldt University of Berlin, Germany
Anders Schoubye, University of Edinburgh, United Kingdom
Camilla Serck-Hanssen, University of Oslo, Norway
Emily Thomas, Durham University, United Kingdom
Amie Lynn Thomasson, Dartmouth College, USA
Giuliano Torrengo, University of Milan, Italy
Barbara Vetter, Humboldt University of Berlin, Germany
Heinrich Wansing, Ruhr University of Bochum, Germany

Titles in the series include:

Knowledge and the Philosophy of Number, by Keith Hossack
Names and Context, by Dolf Rami
The Metaphysics of Contingency, by Ferenc Huoranszki

The Metaphysics of Contingency

A Theory of Objects' Abilities and Dispositions

Ferenc Huoranszki

BLOOMSBURY ACADEMIC
LONDON • NEW YORK • OXFORD • NEW DELHI • SYDNEY

BLOOMSBURY ACADEMIC
Bloomsbury Publishing Plc
50 Bedford Square, London, WC1B 3DP, UK
1385 Broadway, New York, NY 10018, USA
29 Earlsfort Terrace, Dublin 2, Ireland

BLOOMSBURY, BLOOMSBURY ACADEMIC and the Diana logo are trademarks of
Bloomsbury Publishing Plc

First published in Great Britain 2022
This paperback edition published 2024

Copyright © Ferenc Huoranszki, 2022

Ferenc Huoranszki has asserted his right under the Copyright, Designs and
Patents Act, 1988, to be identified as Author of this work.

Series design by Louise Dugdale
Cover image © Temur Pulatov / iStock

All rights reserved. No part of this publication may be reproduced or transmitted
in any form or by any means, electronic or mechanical, including photocopying,
recording, or any information storage or retrieval system, without prior
permission in writing from the publishers.

Bloomsbury Publishing Plc does not have any control over, or responsibility for, any
third-party websites referred to or in this book. All internet addresses given in this
book were correct at the time of going to press. The author and publisher regret any
inconvenience caused if addresses have changed or sites have ceased to exist,
but can accept no responsibility for any such changes.

A catalogue record for this book is available from the British Library.

A catalog record for this book is available from the Library of Congress.

ISBN: HB: 978-1-3502-7714-4
PB: 978-1-3502-7718-2
ePDF: 978-1-3502-7715-1
eBook: 978-1-3502-7716-8

Series: Mind, Meaning and Metaphysics

Typeset by Deanta Global Publishing Services, Chennai, India

To find out more about our authors and books visit www.bloomsbury.com and
sign up for our newsletters.

For Anna and Miksa

Contents

List of Tables	ix
Preface	x

1 The varieties of modal properties — 1
  Possibilities, worlds, contingency — 2
  Dispositions, powers and laws — 5
  Abilities, possibilities and abstraction — 12
  Modal properties, intentional states and realism — 15
  Identification, conditionals and reasoning — 19
  Manifestations, events and actuality — 22
  Contexts, method and ground — 24

2 Dispositions and abilities — 33
  Introducing dispositions — 34
  Being disposed to and being able to — 37
  Why abilities are not dispositions — 40
  Why dispositions are not abilities — 48
  Dispositions and laws — 53
  Teleological dispositions — 59

3 Specificity and extrinsicness — 63
  Specificity — 64
  Extrinsic abilities and abstraction — 72
  Intrinsicness and duplications — 75
  The sense of sensitivity — 80
  Intrinsicness and systems — 85

4 Modal properties and conditionals: The nature of the connection — 91
  The purpose of a conditional analysis: Phenomenalism, ontological reduction or enhanced understanding — 92
  The traditional conditional analysis of dispositions and the standard objections — 96
  Masks, antidotes and circumstances — 101
  Why finks and masks are different — 104

	Losing and gaining abilities	107
	A nonreductive conditional analysis of abilities	111
	'Achilles heels', accidents and the conditional analysis of dispositions	115
5	**Modal properties and conditionals: The scope of the connection**	**125**
	A modal-property analysis of counterfactuals	125
	Some remarks on the logic of counterfactual conditionals	129
	Abilities and nondeterministic processes	133
	Ascribing abilities and reasoning with possibilities	140
	Interactions and the limits of generality	143
	Modal properties without conditionals	148
6	**Manifestations, events and causes**	**155**
	Manifestations and mimics	156
	Events, states and abilities	162
	Manifestations and effects	167
	Causal consequences versus abilities' results	171
	Counting events and abilities	176
7	**Possibility, actuality and worlds**	**183**
	From events to worlds	186
	Contingency and contradiction	189
	Recombination and independence	192
	Essentialism: A world without contingency	198
	The constituents of actuality	203
	Idealism	212
	The system of beings	214
References		**219**
Index		**229**

Tables

4.1   Summary of the objections to TCA   100

Preface

Some things happen or exist only contingently: although they do happen or exist, they do not have to. Some other things do not happen or come to exist, although they could. They are contingent possibilities. Philosophers have tried to understand contingent possibilities in two different ways. According to one, possibilities should be understood with reference to worlds. A nonactual event is possible because there is a world in which it does happen. According to another, possibilities should be understood with reference to modal properties. An event or an object is possible because it would occur or come to exist if a modal property of some substance, another object or set of objects were exercised.

This book argues that the latter approach provides a better account of possibilities than the former. It develops a theory of contingent possibilities according to which possibilities within a world must be distinguished from the possibility of worlds. It is the distribution of modal properties within a world which metaphysically determines what could happen in that world. The most important of these modal properties are abilities and dispositions. Dispositions explain nomic and behavioural regularities, while abilities ground more or less abstract possibilities. By way of connecting the ascription of more or less specific modal properties to different counterfactual conditionals, we can identify various possibilities that we are concerned with when we try to understand what can happen in our world.

During the last couple of years when this book has been written many people helped me in my work. First of all, I wish to thank my colleagues, each of whom is also a friend, at the Philosophy Department of Central European University. I consider myself very lucky that I can teach and think about philosophy in this department's friendly and intellectually stimulating atmosphere. I also wish to thank my students for challenging me on views I thought I properly understood when in fact I just failed to see the difficulties.

I also owe special thanks to Colleen Coalter at Bloomsbury, for making the publication of this book possible, and to Christopher Gauker, one of the editors of Bloomsbury's series *Mind, Meaning and Metaphysics*, for his unfailing support of the project. I am also grateful to my anonymous referees for their helpful comments on the penultimate version of the manuscript.

Finally, I wish to acknowledge the generous support of Central European University which granted me a sabbatical leave and financial support to complete this project. CEU has gone through very hard times recently. It was founded, and it has grown to be an internationally renowned centre of both study and research, in my homeland, Hungary. But, while I was writing this book, the university was forced to leave the country by a corrupt, autocratic regime operating on the basis of a dark ideology, which openly rejects the values of enlightenment and rationality, in the twenty-first century, in the middle of Europe. Our university moved to Vienna. So, lastly, I would like to thank my colleagues in Austria for welcoming us so warmly and opening many new possibilities when some others have been lost.

1

The varieties of modal properties

One of the most influential metaphysicians of the twentieth century, David Lewis, said once of events that they are not much of a topic in their own right.[1] The same seems to have been true for a long while about dispositions and abilities. Most philosophers in the early twentieth century were interested in dispositions – and, to a lesser extent, in abilities – because of the role they play in philosophy of science, philosophy of mind, philosophy of action or moral psychology. And, as the spirit of the age demanded, the problem of dispositions was mainly discussed in the context of philosophical semantics. Philosophers were most interested in the analysis of the meaning of disposition *terms*.

But the situation changed significantly by the end of the twentieth century. By then, understanding dispositions became a central issue in the philosophy of science. Dispositions occupied a particularly important place in the debates over the metaphysical grounds of natural laws and causation. New realists about dispositions challenged the Humean consensus, according to which there are not 'necessary connections' in nature; or more specifically, the 'dogma' that laws of nature and causation are metaphysically contingent. Dispositions have been reinterpreted as causal powers which can ground the necessity of laws and causal connections.

The purpose of this book is to reconsider the importance of dispositions and abilities from another perspective. I agree with the new realists that Humean scepticism about modal properties is unmotivated and that the philosophical problem of modal properties should not be interpreted as an attempt to provide a 'reductive analysis' of disposition terms. I also agree that dispositions must play a foundational role in any plausible account of the nature of natural laws. However, I am not convinced that dispositions' role is to ground any necessity in nature or that laws of nature would entail any kind of necessity.

---

[1] See Lewis (1986b, 241).

Put otherwise, one of my central claims in this book is that, if causal powers do indeed ground 'necessary connections in nature' – as the Humean tradition, which associates the attribution of powers with necessity, has it – then dispositions that ground laws are *not* causal powers. Much more importantly, I shall argue that we need to distinguish a type of modal properties which I propose to call *abilities* from modal properties that are commonly called dispositions. That distinction plays a crucial role in the project of this book, which is to develop a theory of contingent possibilities that is grounded in an analysis of abilities.

Possibilities, worlds, contingency

The idea that contingent possibilities should be understood with the help of modal properties originates in the Aristotelian metaphysical tradition. According to that tradition, what can happen in the world is grounded in properties instantiated by things in it: substances, persons, objects or the systems which are composed by them. Aristotle's word for these properties is *dunamis*. We can find in Aristotle's notion of *dunamis* the origin of both kinds of modal properties that are relevant to the topic of the present work. In one sense, Aristotle's *dunamis* is the ground of behavioural tendencies – what happens regularly, but not always. In another, it is the ground of possibilities, some of which are actualized at a certain place and time, while others remain unactualized then and there.[2]

The currently dominant understanding of possibility is the Humean one. Humean accounts of modality are based on the rejection of Aristotelian ontology, which assigns a crucial role to modal properties. Humeans reject first-order modal properties. Instead, they try to understand possibilities in terms of *worlds*. According to their view, 'An object can do x' should be understood as 'There is a possible world in which the object or its counterpart does x'. In contrast, according to the account of the present work, which is broadly Aristotelian, 'An object can do x' should be understood in terms its ability to do x in this world.

---

[2] See, respectively, Hintikka 1973 and Beere 2009. I shall only rarely use the notion of 'potentiality'. Some philosophers use potentiality as the ground of natural necessity, others as the ground of possibility. Historically, potentiality does not express a property of an object, but rather *a way of being*: something can exist or happen potentially and actually. 'Potential' has also very specific meanings in physics and in neuroscience. Thus, although 'potentia' is the Latin term for the Greek 'dunamis' which expresses, among other things, the modal properties of objects that ground contingent possibilities, the use of that word can have many, as we shall see occasionally misleading, connotations. In the next section I shall suggest using 'potentialities' as expressing possible events rather than describing the properties which ground them.

Why do the Humeans want to understand possibilities with reference to nonactual *worlds*? I shall return to this question in the final chapter of the book. I shall suggest there that the world-account of possibility rests on the assumption that the model of contingency should be *logical* possibility. Something is logically possible if it does not involve a contradiction. But I shall also argue that logical possibility is not a good model for understanding contingency. For logical possibility is not incompatible with necessity, but contingency is.

If something is logically necessary, it is also logically possible. In a necessitarian world – the world of Spinoza's and perhaps Leibniz's – everything happens by necessity, but many things are *logically* possible: everything which actually happens. However, in a necessitarian world, there is no contingency since, by definition, the necessity of events entails their non-contingency. What is possible in the sense of being contingent cannot be necessary.

Despite this obvious difference between logical possibility and possibility in the sense of contingency, most modern philosophers try to understand possibility on the model of logical possibility. According to their view, something is possible only if it does not involve a contradiction. Trivially, a necessary truth does not entail any contradiction; hence, what is necessary is also possible.

More importantly, contradiction, in its primary sense, requires propositions. Philosophers often talk about 'conceptual impossibility'. But if conceptual impossibility is understood as a contradiction, then conceptual impossibility must presuppose contradictory propositions. If a 'married bachelor' is a 'conceptual impossibility', it is that because 'S is bachelor' and 'S is married' are contradictory propositions; it is in this sense that the statement 'S is a married bachelor' must 'involve a contradiction'.

Why is this important for our philosophical understanding of modality? Because only a *set* of propositions can be contradictory. Logical possibility/impossibility is tied to the notion of consistency, and consistency presupposes a set of propositions. This is, I suggest, the origin of the world-account of possibilities. According to the world-account of contingency, a proposition *p* can be true if and only if *p* is possibly true in the sense that there is a possible world in which *p* is true.

However, this approach is beset with a big problem: what determines whether a proposition can be true in *some* possible world? Obviously, it is how properties can be distributed in a world. But then that which grounds possibilities must ultimately be explained by the nature of properties *in* a world, not by the existence of other worlds. Thus, following the Aristotelian tradition, in order to understand the nature of contingency we have to give an account of the modal properties instantiated in this world, rather than trying to understand it with reference to others.

My preferred account of contingent possibilities presupposes realism about modal properties. By realism, I mean that such properties are (*i*) mind-independent and that they are (*ii*) first-order properties of persons, substances, objects or their ensembles, not just second-order surrogates of first-order nonmodal ('categorical') properties. Realism thus entails that (*i*) modal properties are not merely projected onto the physical or material world as Hume – at least according to the traditional interpretation of his views – thought they are, and that (*ii*) modal properties should not be understood as resultant properties to which some nonmodal properties, together perhaps with some laws, give rise.

One important consequence of this view about modal properties is that modal truth cannot be understood in terms of how the assumedly nonmodal properties are distributed in the world. As mentioned, most contemporary accounts of modality presume that modal truth should be explained in terms of worlds, and worlds should be identified with reference to the distribution of nonmodal properties, often called 'categorical properties' or 'qualities'. This book is motivated by the thought that we need an alternative account of the nature of modal truth.

My rejection of the world-based account of modality is not new. There have always been philosophers who were sceptical about the world-based account of possibility (or modality in general). However, most of them were sceptical about the world-based account of modality because they had problems with the very idea of possible worlds. Though the concept of possible worlds is routinely used by many contemporary philosophers in several contexts, other philosophers have found this concept unintelligible.[3]

I do not have any objection against the idea of possible worlds as such. If we grant, as in my opinion we should, that our world is a contingent one, then we could not deny that there are worlds that are 'merely' possible (that is to say, nonactual). However, not finding the notion of possible world unintelligible is one thing; understanding possibilities in *this* world in terms of what happens in others is another. The world-based accounts interpret possibilities in this world with reference to other possible worlds. According to those accounts '*a* is possibly F' or '*a* is possibly φ-ing' just means that there is a possible world in which *a* (or its counterpart) is F or *a* (or its counterpart) is φ-ing. It is this understanding of

---

[3] For an early summary of such worries see Mondadory and Morton (1979). For an account of modality which is motivated by such worries see Jacobs (2010).

possibilities in our world which I find unilluminating and which, I think, should be replaced by an alternative account.

Dispositions, powers and laws

For this project to be carried out, we need first to introduce some distinctions within the class of modal properties. Modal properties are *dynamical* properties. This means that, as Aristotle thought, they are the origins of changes in the world. As we shall see – and as Aristotle mentions too – this is a bit of simplification, because dynamical properties can also be the origins of resistance to certain kind of changes, or other sorts of non-changes like persistence and equilibria resulting from the joint exercise of different modal properties. Thus, more generally, modal properties are the sources of both events and states.

However, modal properties are not only the sources of actual events and states. They are also the grounds of possibilities. Events happen when modal properties are exercised, but most of these properties remain unexercised during most of the time. In current discussions, 'dispositions' are often taken to be synonymous with 'modal properties' that are the sources of changes. But dispositions are just one species within the larger class of such properties. They need to be distinguished from other types of modal properties like old-school 'causal powers' and, even more importantly for the project of this book, from properties which I propose to call *abilities*.

It might seem then that, initially, we need to start with a general account of what makes a property *modal*; that is, an account of how such properties are different from the assumedly nonmodal ones. In fact, however, this is the last question we need to answer. It is often assumed without much argument that we clearly understand the so-called nonmodal or 'categorical' properties, while the possibility of modal properties needs some further explanation. On reflection, however, as I shall argue in the concluding chapter, it may turn out that precisely the opposite is true. We cannot understand the nature of any property without a prior understanding of the modal ones.

The *dispositional* versus *categorical* distinction was introduced by Gilbert Ryle.[4] But in current discussions about the metaphysics of properties, it is often ignored that Ryle introduced this distinction in order to distinguish certain

---

[4] See Ryle (1949, 117); and further in Chapter V, section 2 with the title 'The Logic of Dispositional Statements'.

types of *statements*, not to classify properties. His claim was, roughly, that, sentences which ascribe dispositions to objects are not fact-reporting categorical statements but condensed hypotheticals. The statement 'This glass is fragile' may look as categorical, fact-reporting as, for instance, 'This glass is on the table' is. But, on Ryle's view, the former statement does not report any fact. Rather, what it means is something *hypothetical* like 'This glass would break, if it fell from the table'. His account of what is dispositional and what is categorical was not meant to characterize properties but *statements* with reference to certain types of terms figuring in them.

In fact, I have doubts that the very idea of 'categorical property' as it is standardly understood in contemporary metaphysics has a clear meaning or that it can do much work in ontology. But, as said, I can deploy the reasons for my doubts in detail only after I will have provided an account of the nature of modal properties. The point that I wish to stress already at this introductory stage is only that we cannot provide a proper account of modal truth without introducing some more fine-grained distinctions within the larger set of modal properties. First, we need to distinguish (*i*) old-school causal powers from dispositions, and then, more importantly for the project of the current book, (*ii*) dispositions from abilities.

New realist accounts of modal properties often identify dispositions with causal powers, which are assumed to ground (non-Humean) natural necessities. Natural necessities then may or may not be understood as metaphysical necessities. But according to my account, while dispositions do indeed ground laws, this does not mean that laws of nature would involve any 'necessary connection'.[5]

Since many laws are, or at least entail, regularities, if laws expressed *de re* necessities, it should be true that the occurrence of certain nomological conditions C invariably correlate with some nomological consequences N. In fact, however, most nomological generalizations are true only *ceteris paribus*. And the few exceptionless laws with which we are familiar (for instance, laws of

---

[5] For powers involving a special kind of necessity, see Harré and Madden (1975); for powers entailing metaphysically necessary truth, see Ellis (1999) and Bird (2007a). In a broad sense, my account is akin to Anjum's and Mumford's in Anjum and Mumford (2018), though it is different from theirs in many details. As it will become clear soon, I take dispositions to ground laws and not causation, and I understand tendency as a statistical notion. Most importantly, I see a tight connection between the ascription of modal properties and conditionals. In my earlier publication on this subject (Huoranszki 2012), I called 'powers' what I call abilities now because I restricted the ascription of abilities to properties of animated beings. My arguments in Chapter 2 will make it clear why I have changed my mind on that issue.

conservation or some exclusion laws) can hardly be explained in terms of *causal* powers.

Nonetheless, all laws, exceptionless or not, are grounded in dispositions. Some laws, as I shall argue in Chapter 3, are grounded in individual objects' (or particles') dispositions; others – like most physical laws – are grounded in a system's dispositions. But such dispositions are not to be interpreted as causal powers.

If we follow Hume and the tradition established by him, then we need to admit that the ascription of causal powers must entail a commitment to 'necessary connections' in nature. But while nomic connections do indeed have a special modal status in the sense that classifying a regularity as 'lawlike' involves a commitment about its role in grounding some modal truth, the modality involved is not a necessity. It is generally agreed that if the occurrence of an event or a state is a consequence of some nomic regularities and conditions, then there must be a sense in which it is not an accident. I shall argue that the sense in which it is not an accident can be best explained in terms of *nomic dispositions* which ground the nomic connections between the occurrences of certain states and events.

Although non-accidentality is a standard feature of the nomic, its precise meaning has not been properly explained so far. In Chapters 2 and 4, I shall argue that the relevant sense of non-accidentality should be cashed out in terms of the non-accidentality of nomic consequences rather than in terms of non-accidentality of laws themselves. And further, I shall argue, that the relevant features of non-accidental connections derive from the nature of dispositions, which, together with natural kinds and physical kinds, grounds those laws.

Another important reason to distinguish dispositions from powers is that the notion of power is commonly associated with causation. Powers are meant to be *causal* powers. But most dispositions need not (and often cannot) be causal. Causation is a form of interaction. But many laws which dispositions ground are not laws of interactions. They are laws about instantaneous states of a system or about processes and propensities of its evolution. Moreover, even when dispositions do ground laws of interactions, such interactions need not be causal.

One reason why many philosophers take dispositions to be causal powers is that they are afraid that, unless we do so, we must take dispositions to be 'causally impotent'. However, as I shall argue in Chapter 6, this worry seems to me entirely misplaced. For in *one* sense not only dispositions, but also powers, are 'impotent'. And in *another* sense, every modal property as a real property of some object(s) must be 'potent', irrespective of whether or not it is 'causal'.

In one sense, dispositions and powers are not potent simply because they are properties, and it is substances, objects or their systems, and not their properties, that can *have* potencies. Suppose that dispositions and powers themselves could be 'potent' (in contrast to being 'impotent'). Then such potency should be a second-order property of some first-order modal properties. But there is no such second-order property. Modal properties *specify* objects' potencies: *objects* have potencies in virtue of their modal properties. For modal properties, strictly speaking, do not *do* anything: it is always objects (chemical substances, systems of particles, etc.) having the properties which act, react or interact.

In fact, if we understood potency as a second-order modal property of first-order modal properties, a regress would immediately be kicked off. For it may then be asked: what makes this second-order modal property 'potent'? Following this logic, we should say that only a property of an even higher order can do so: the potency of the potency of the first-order property. And so on. Some consider this as a challenge to the very idea of first-order modal properties. But the regress induced by this reasoning seems to me entirely bogus. It is created by the unnatural assumption that modal properties themselves, rather than the objects having them, possess 'potencies'.[6]

In another sense, however, dispositions might be understood as *being* – not *having* – 'potencies'. If we accept the Eleatic Visitor's dictum in Plato's *Sophist* according to which 'those *which are* amount to nothing other than *capacity* (*dunamis*)' and, as in Plato, we take such capacities as dispositions which ground interactions, then many dispositions are certainly potencies.[7] For, as we shall see, dispositions presuppose abilities and many abilities ground the ways in which their bearers typically interact with other objects or substances. Whether or not those activities and interactions are *causal* in any sense in which causation is used in contemporary philosophy is, however, a further question.

According to the standard contemporary understanding, interactions are causal only if they are asymmetric (causes are not the direct effects of what they cause). But most nomic interactions grounded in dispositions are perfectly symmetric. My claim is that we need to distinguish causal powers from dispositions because causal powers must ground *asymmetric* interactions between objects. However, most dispositions ground symmetric interactions in the sense that it is impossible for an object participating in the interaction

---

[6] Psillos advances a similar argument about powers when he claims that unmanifested powers must have the power to manifest themselves (Psillos 2006, 19). For an Aristotelian response see Marmodoro (2010a).
[7] Plato, *Sophist* 247e.

to exercise its disposition without the other participating object's exercising a corresponding disposition.[8]

To invoke Malebranche's standard example of mechanical interaction (made popular among empiricists by Hume), when two billiard balls collide, both are disposed to change the momentum of the other according to the laws of collision of rigid bodies. But, contrary to Hume, there is no objective reason to call the change of momentum of one ball as a 'cause' and the change of momentum of the other as an 'effect'. 'Constant conjunction' might be turned into an asymmetric correlation by temporally arranging the conjuncts, but typical physical interactions between bodies involving a change in their states are simultaneous.[9]

Given the asymmetry of causation, dispositions can be causal only when they ground asymmetric interactions. An asymmetry might occur when an interaction involves changes only in one of the interacting objects (or substance, or system of objects). An example might be placing a magnet near to a compass. The presence of magnet causes changes in the position of the compass's needle without the magnet (seemingly) undergoing any change as a response to the position of the needle.[10]

Modal properties which ground such 'asymmetric interactions' might be considered as causal powers. In fact, it seems, that it was this type of 'interaction' that once served as a model for power-ascriptions: God has the power to change anything without himself changing at all. But it is important to note that in the created world such causal powers can be ascribed, if they can be ascribed at all, only to macroscopic objects (or to persons). All nomic interactions at the microphysical particle level are perfectly symmetrical. This seems to me a

---

[8] The most obvious case in point is, of course, physical bodies' interaction when they obey Newton's third law of motion.
[9] A possible response to this is to claim that causation must be, after all, simultaneous. This is the view of Mumford and Anjum (2011), Chapter 5. But while I think that their 'vector-theory' presents an interesting and original view of the nature of (at least some, if certainly not all) nomic interactions, it seems to me to be so far from what we ordinarily and traditionally think about causal connections, that I have difficulty with understanding it as an account of causation. For further considerations of these lines see Glynn (2012) and Bird (2016).
[10] In distinguishing causations and causal powers from modal properties, the exercise of which involves simultaneous interactions, I follow Kant (and Hegel, who himself seems to follow Kant in this regard). In the Analogies of *The Critique of Pure Reason* Kant distinguishes reciprocal interaction from causation. This is not an arbitrary terminological distinction, but an observation of great metaphysical significance, since, for Kant, laws of reciprocal interactions are metaphysically connected to simultaneity and persistence, while causation and causal laws are connected to succession and alteration. But we need not follow Kant in every respect in order to agree with him in that reciprocal interactions and causations are two metaphysically different categories.

further reason to view causation as an emergent, macroscopic and perhaps even mind-dependent phenomenon.[11]

Be this as it may, what matters for us here is that even if there are 'objective', mind-independent causal powers which can ground some asymmetric interactions, the concept of dispositions must be understood independently of the study of causation. This does not mean that the ascription of dispositions cannot play some derivative role in an account of causation. But it does mean that dispositions need to be distinguished from causal powers. And, even more importantly, it means that we should not try to understand dispositions in terms of causation. In Chapter 6, I shall explain in more detail why dispositions (and abilities) cannot be analysed with the help of causal processes as some currently popular accounts try to do.

Some might complain that if modal properties are not powers and hence not causal, such properties cannot *explain* the behaviour and interactions of objects (or particles, substances, etc.). However, explanatory irrelevance does not at all follow from the absence of the use of causal language. The idea that every good explanation of contingent events must be an explanation by their causes is another legacy of Hume's philosophy. But realists about modal properties need not, and if I am right, should not, accept this legacy.

Even if we set aside the issue of irreducibly functional/teleological explanations in the life sciences, we should observe – as Russell famously argued[12] – that most nomological explanations in physics are not causal. Hence, if many dispositions ground physical laws as they do, the way they contribute to the explanation of physical phenomena cannot be causal. Further, as Ryle notes, we can often explain events by ascribing dispositions to objects better than by identifying their causes. In many contexts, we can better explain the breaking of the glass with reference to its fragility than to the event of something hitting it. For, if two glasses are similarly hit by a stone and one breaks while the other does not, it is the difference in their fragility what best explains what happened. But this does not imply that we should say that the fragility, rather than the stone or its trajectory, 'caused' the breaking.[13]

Even more importantly, as I shall argue in Chapter 6, there is a special way in which modal properties explain. We may call that form of explanation

---

[11] About causation as being emergent see, most importantly, the collection edited by Coary and Price (2007).
[12] In Russell (1976/1912).
[13] Thus, contrary to what is often assumed, Ryle did take disposition terms genuinely explanatory. He says, 'The explanation is not of the type "the glass broke because a stone hit it", but more nearly of a different type "the glass broke when the stone hit it, because it was brittle." Ryle (1949, 50).

'explanation by identification'. As I shall show there, we often identify processes or events with reference to the modal property which they manifest and thereby we can also answer the question about why they have occurred.[14] In general, I suggest avoiding the widely shared custom in contemporary philosophy to identify causation with what can be explanatorily relevant.[15]

Finally, the notion of dispositions is often mentioned in the context of ascribing habits and personality or character traits to persons. In fact, arguably, this was the original meaning of the word 'disposition' in English. And although habitual behaviour obviously manifests an agent's habits, it is hard to make sense of the idea that a habit causes any of their actions. Smoking a cigar can – though need not – show that an agent is a smoker, but his being a smoker does not cause his smoking-behaviour on any specific occasion.

Despite this similarity in their explanatory potential, and despite considerations about the origin of the use of the term, I shall argue that we must clearly distinguish dispositions as habits or character traits and dispositions as nomological properties. The ontological role of the latter is to ground, and hence explain, the special modality of laws. The role of the former is to ascribe a property by which we can make sense of certain forms of behaviour. But this does not entail any special commitment about the modality of the specific behaviour. The explanation of behaviour with reference to habits does not entail the non-contingency of an action in any decent sense.

There are a number of differences between these two uses of 'disposition' which I shall discuss later in Chapter 3, but there is one that I would like to call attention to already here. If habits are understood as 'second natures', then they must be acquired dispositions. But a habit's acquisition is different from the ways in which other modal properties are acquired. It is not to be understood on the model of such cases as when a magnetizable object becomes magnetic; that is to say, when an inanimate object has the disposition to acquire another disposition.

A nomological process often (perhaps always) terminates in the change of some object's disposition(s). A change of a body's momentum involves

---

[14] This sort of explanation is similar to that which was once called, a bit misleadingly, an 'explanation by redescription' in action theory.
[15] It has become a convention to translate, following the standard Latin translation of his works, Aristotle's notion of 'aitia' as the 'cause' of what is explained. But for Aristotle, *aitia* is not a cause in the modern, post-Cartesian sense. For Aristotle, anything that provides an answer to a why-question is an 'aitia'. (See *Physics* III.2 and *Metaphysics*, Book Delta, II.) Hence, it would be clearer to understand what is often called the 'four causes' as four types of *explanatory factor*. Interestingly, such explanatory factors themselves are not called 'aitia' by Aristotle. What is translated as 'final cause' is, for instance, 'hou heneka', that is, 'what something is for' or 'for the sake of which'; efficient cause is 'that whereby the motion/change start' or (in which it originates). For a brief, illuminating and authoritative discussion of this issue see Moravcsik (1974, 6–9).

changes in its dispositional properties (like its kinetic or potential energy). Such changes must obviously be distinguished from the changes that happen when someone acquires a habit, which can hardly occur as a result of a single interaction. 'Acquisition' in the context of habitual dispositions involves learning or regular exercise; and, unlike in the magnetic-magnetizable case, the acquired dispositions are often, though not always, teleological.

Abilities, possibilities and abstraction

My main concern in this book is not, however, dispositions, habits and laws, but abilities. Dispositions need to be distinguished not only from causal powers, but also, and more importantly, from abilities. While the ontological role of dispositions is to ground nomic regularities and habitual behaviour, abilities' role is to ground possibilities. My most important aim is to explain why understanding abilities as a distinctive type of modal properties can provide a better account of the nature of contingent possibilities than traditional 'world-based' accounts do.[16]

As said, I shall offer a property-based account of contingent possibilities. If possibilities are meant to express contingencies, then they cannot be grounded in some *a priori* logical or mathematical truth or principle. They must be grounded in the modal properties which objects, persons, chemical substances instantiate. And the relevant modal properties which ground such possibilities are neither causal powers nor dispositions, but persons', objects' or substances' *abilities*. Since the distinction, as well as the connection, between abilities and dispositions plays a pivotal role in my account of modality, I devote the whole Chapter 2 to explaining them.

I am going to use four different kinds of argument to justify the fundamental difference between properties that are abilities and properties that are dispositions. Some of the arguments are based on considerations about the history of these concepts, that is, how they have been applied in philosophy. Others derive from observations about how we ascribe abilities and dispositions in English. More specifically, I shall argue that '... is able to do ...' is semantically connected to '... can/could do ...', while '... is disposed to do ...' is connected to

---

[16] Similar ideas can be found in Williams and Borghini (2008), Jacobs (2010), Vetter (2015) and Pawl (2017). The idea that our ordinary use of can-statements in the sense of *de re* possibilities are grounded in our practice to ascribe modal properties to objects first appears in contemporary analytic philosophy in Wiggins (2001, 112–13).

'... has a (certain statistical) tendency to do ...'. Further, I shall also try to bring out the different logical implications of statements ascribing abilities *versus* statements ascribing dispositions.

The most important reason for distinguishing abilities from dispositions is, however, their different role in ontology and the different ways in which we apply them in our modal reasoning. The most pressing reason to introduce abilities as a special type of modal properties is not historical, semantic or logical, but ontological. Abilities, I shall argue, play a prominent role in ontology: they ground contingent possibilities, and they also ground the ways we reason with such possibilities in both practical and theoretical contexts; while, as we have seen, the ontological role of dispositions is to ground nomological and habitual behaviour.

A thing might have the ability to do something only once and in very specific circumstances. Given the nature of 'butterfly effects' (in chaos theory), one might be *able* to (and hence can), just once and for a short moment, break faraway windows by a sneeze.[17] It would be absurd to claim though that one is *disposed* (and hence has the nomic tendency or a habit) to break faraway windows in the sense in which one might be disposed (and hence has the nomic tendency) to infect bacterially others in one's vicinity by a sneeze.

At this point, another distinction between dispositions and abilities becomes crucial. Abilities, unlike dispositions, can be maximally specific. I have the specific ability to type the word 'sneeze' *here and now*, and you can justifiably ascribe that ability to me, even if you cannot explicitly specify every condition which needs to be satisfied in order for the word 'sneeze' to be typed by me. But I also have the very generic ability to break faraway windows by a sneeze in circumstances which almost certainly do not occur here and now (and probably never will occur during my lifetime), but which you can justifiably ascribe to me if you can explicitly specify all those conditions which need to be satisfied for me to perform that feat. The possibility that this generic ability grounds is an uninterestingly faraway/abstract possibility, indeed; but it is a possibility, nonetheless.

Those who deny the existence of first-order modal properties usually hold that objects' abilities should be explained with reference to their behaviour in other possible worlds, or that actual behavioural tendencies ground the *ascription* of disposition *terms* (not so much dispositions themselves, which might not even

---

[17] To use Alexander Bird's very telling example, to which I shall return in more detail later in Chapter 3. See Bird (1998).

be, on those views, *sui generis* properties). I shall challenge such views and argue that (*i*) the metaphysical relation is the opposite in the case of dispositions: it is substances', objects' or their system's dispositional properties that ground nomic or habitual tendencies in the world; and that (*ii*) the metaphysical relation is entirely different in the case of abilities, since abilities ground contingent possibilities and hence they are not to be understood in terms of possible worlds at all.

As the example of 'butterfly effect' illustrates, intuitively, possibilities could be more or less 'close' to actuality. Breaking windows by a sneeze is a 'faraway' possibility, while breaking a vase by not handling it with care is a 'close' one. According to the world-based approach, such metaphors about distance should be taken almost literally: we distinguish possibilities by 'how distant' the worlds in which they are true are from our actuality. Unfolding this metaphor, the world-account relies on the idea of 'comparison' of worlds: 'distance' is determined by 'how similar' other worlds are to the actual.

In Chapter 3 I shall suggest that, rather than trying to interpret intuitive differences in possibilities in terms of 'distance' and 'similarities of worlds', we should understand them in terms of their abstractness. I shall argue that possibilities are abstractions from the concrete actuality and abstractness can come in degrees. And further, the degrees of abstractness of contingent possibilities are grounded in the specificity and generality of things' abilities.

As we shall see, the issue of specificity and generality is essentially connected to the problem of intrinsicness of the abilities which ground possibilities. Intrinsicness, like specificity and abstractness, is a matter of degree. Suppose a glass is filled with some protective material. Is it able (or liable) to break when dropped? No, it is not in the concrete, actual circumstances. But in the same circumstances, it does have the ability to break when, after the removal of protective material, it is dropped. More generally, something which lacks a *more determinate* ability to φ at time t in the actual circumstances C (for instance, the glass in our example which is actually filled with protective material at t would not break if dropped) can have at the same time *a more generic* ability to φ (for instance, the same glass would break if it were dropped, *and* the protective material were removed from it).

Concrete, determinate events occur only when and where abilities are exercised. Unexercised abilities ground unrealized possibilities, which are necessarily abstract and indeterminate *to a certain extent*. But abilities can be very generic and very specific in the sense indicated above. Correspondingly, the possibilities that they ground can be more or less abstract. How abstract a possibility is depends on 'how far' it is from the concrete reality; and how far it is from concrete reality is determined by how many unrealized conditions need to be realized in order for an ability to be exercised. The fewer conditions of an ability's exercise are actually satisfied, the more abstract the possibility grounded by it is. This tight connection between (*i*) the abstractness of possibilities (i.e. 'how far' they are from concrete reality) and (*ii*) how specific or generic the abilities which ground them are is the essential idea on which my account of contingency rests.

Modal properties, intentional states and realism

Specificity and generality, intrinsicness and extrinsicness, concreteness and abstractness are ontological distinctions. A closely related semantic and epistemic issue about abilities concerns the ways in which we can *specify* them. It is obvious that there are many more abilities than 'disposition terms', or *words* that we can use in normal contexts to express a modal property. Hence, we cannot really express every ontologically relevant distinction among abilities with the application of such terms. How can then we identify the abilities which ground the corresponding possibilities?

Some philosophers hope to answer the problem of specification by trying to exploit a similarity between ascriptions of dispositional properties and ascriptions of states of mind.[18] Most, perhaps all, mental states are intentional in the sense that they are directed at some object. We cannot just think without thinking about something; and we cannot just see without seeing something. Thoughts and perceptions are always *about* some object. But an agent can perceive, or can think about, an object which does not actually exist. One can think of an elephant that is as small as a mouse and one can perceive (when hallucinating) an elephant which is pink throughout.

It has been argued that dynamical-modal properties display a similar feature and that this is sufficient for identifying them. An object's fragility is directed at breaking, even if the object never breaks. Modal properties are 'directed at' certain events. Those events need not actually occur, but the relevant properties can be distinguished and specified with reference to them.

---

[18] These include Place (1996), Martin and Pfeifer (1986), Molnar (2003), and Lowe (2010). The idea that powers are like mental properties played an important role in early modern arguments against 'occult qualities', that is, properties that we would call today 'dispositional'. About this interesting historical issue see Ott (2013), especially Chapter 3.

The key idea is that, just as mental states may be identified with reference to an intentional object which might not exist, dynamical properties may be identified with reference to events which perhaps never occur; but which would occur if the property were exercised. This does not mean that all modal properties are conceived to be mental. What it means is that not all intentional states are mental. A fragile object can break, and a soluble substance can dissolve. In this sense, fragility is directed at breaking, solubility is directed at dissolving.

The intentionalist view might be correct as far as very generic *capacities* and *know-hows* are concerned. And, no doubt, the ascription of such capacities can play some role in scientific theorizing as well as in our understanding of human and animal behaviour. However, in many contexts, they are not the sort of properties in which we are interested when we ask what can happen or could exist. I might have the generic capacity to swim in the sense that I know how to swim. Nonetheless, I cannot swim if my legs are broken.

More importantly, such 'intentional objects' cannot discriminate abilities properly. A piece of litmus paper can turn red, and it can turn blue. According to the intentionalist account then, it should have the capacity to turn red and the capacity to turn blue. But why does it not have the capacity to turn red or blue? Because its turning red and its turning blue manifest two distinct, more specific, abilities. Litmus papers are useful for chemists precisely because they have these more specific abilities, not just some generic capacities of changing their colour to red or to blue; a very boring capacity indeed, which basically all macroscopic objects possess.

According to the intentionalist view of property-identification, we can identify more specific or finer-grained modal properties with the help of specifying the events at which they are directed more finely. We can say that one does not have the ability [to swim with broken legs]; or that a litmus paper has the ability to turn red in acidic solutions and that it has also a distinct ability to turn blue in alkaline solutions. However, this makes sense only if there are such *events* as 'swimming with broken legs'; 'turning red when immersed in acidic solution' and so on.

In Chapters 5 and 6 I shall argue that we can, in a way, make sense of such events. But we cannot make sense of them *without* the use of counterfactual conditionals. Thus, the specification of certain modal properties must exploit the connection between counterfactual conditionals and the ascription of abilities. Litmus papers are such that they would turn red if they were immersed into acidic solution; and they are such that they would turn blue if they were put in a basic (alkaline) solution. The use of counterfactual conditionals is far the most natural way to specify most abilities. But this means that there must be a conceptual connection between the ascription of a specific ability and the truth of certain counterfactual conditionals.

Many philosophers hold that postulating such conceptual connection is incompatible with realism about first-order modal properties. For the so-called conditional analyses of dispositions, which aim to uncover the exact form of the connection, are often associated (both by its critics and by its advocates) with an attempt to give a reductive account of modal properties. Those who reject first-order modal properties in ontology typically try to understand the ascription of modal properties – most often, objects' dispositions – with the help of counterfactual conditionals. As a seemingly natural reaction, most realists about such properties deny that there is a logical connection between the ascription of modal properties and the truth of counterfactual conditionals. Do I not make a concession to antirealism by requiring an essential connection between the ascription of abilities and the truth of certain conditionals?

In order to answer this question, we first need to clarify what 'reduction' might mean in this context. Antirealists aim to reduce modal properties in two steps. First, they argue that there must be a logical connection between the ascription of a dispositional property and the truth of a counterfactual conditional. Second, they explain the truth or falsity of the relevant counterfactual conditionals with reference to which actual regularities occur in the world or, alternatively, in terms of possible worlds.

Hence, there might seem to be a natural association between the rejection of the Humean view of properties and the possible world account of contingency on the one hand, and the denial of logical connection between the ascription of modal properties and counterfactual conditionals on the other. But this association, even if natural, is misleading. For it is not at all inconsistent to grant the existence of first-order modal properties in ontology *and* to accept that there is an essential connection between the ascription of abilities and the truth of certain conditionals. It is one thing to say that the ground of the truth of some conditionals is the instantiation of certain modal properties; it is quite another to deny that there is any logical connection between the ascription of modal properties and the truth of counterfactual conditionals.

According to the standard Humean approach, the ascription of a modal property entails, and is entailed by, the truth of some counterfactual conditional. Simplifying greatly what will be discussed in more detail later, an object's being fragile entails, and is entailed by, the truth of some statement like 'This object would break, if it were struck'. What distinguishes the original Humean accounts of dispositions from the realist one is that, on the Humean accounts, the ultimate ground of the truth of such counterfactual conditionals is some statistical or nomic regularity. Since it seems indeed to be a nonaccidental and hence nomic regularity that certain kind of objects, when they are struck, break, it follows that an actually unstruck object of that kind would break as well if it were struck.[19] Nomic regularities 'support' the truth of counterfactual conditionals.

More sophisticated Humean accounts of such conditionals analyse them with reference to possible worlds rather than to nomic regularities. According to those accounts, a counterfactual is true if its consequent is true in the most similar, or in most of the most similar, possible world(s) in which its antecedent is true. However, these more sophisticated accounts do not ultimately differ from the earlier ones. For although the 'similarity of worlds' can be measured according to different parameters, it is generally granted that the most important one is the sameness, or 'non-violation', of nomic regularities.[20] Thus, even in this case, the ultimate bases of the ascription of modal properties are those nomic regularities which obtain in the actual world, and the sameness of which determines 'how far' a world is from the actual one.

I shall argue that what distinguishes the realist approach of modal properties from the Humean is not the former's denial of the essential connection between the ascription of modal properties and the truth of certain conditionals, but their rival views of explanatory priority. On the realist account, the ground of the truth of counterfactual conditionals, just as much as the ground of nomic regularities, is the distribution of modal properties in the world. This does not mean, however, that we can give an adequate account of modal properties without exploiting their logical connection to certain conditionals.

Since this is a contentious claim, let me finally mention that, in denying that the ascription of modal properties is logically connected to the truth of conditionals, contemporary realists seem to part company with their most important historical predecessor. I do not mean this as an argument for my own views, but it is interesting to note that, although Aristotle does not, of course, introduce any 'conditional analysis' of capacities, he certainly understood dynamical properties in a way that is compatible with it.

In Book IX, section 5 of *Metaphysics*, Aristotle says that '*when* the agent and the patient meet in the way appropriate to the potentiality in question, [*then*]

---

[19] See Quine (1960, 225). As we shall see later in the next chapter, Ryle compared dispositional ascriptions to nomological regularities.
[20] On the metric of similarity between worlds and on the priority of sameness of nomic regularities among the various dimensions of comparison, see Lewis (1986/79).

the one must act and the other be acted on' (1048ᵃ6-7). And further, he also claims that 'that which is capable is capable of something and some time and in some way – with all other qualifications which *must be present in the definition*' (1048ᵃ1, my emphasis). Capacities (or rather: abilities) are *defined* with reference to the circumstances in which they would be exercised.[21] This is certainly not a conditional analysis of modal properties that ground possibilities, but it is near enough to serve as a starting point for my reinterpretation of its significance.

Of course, my aim in this work is to engage in some contemporary debates, not to provide an interpretation of Aristotle. Aristotle's work is, undoubtedly, an important source of inspiration for those who search for an alternative to the possible world interpretation of contingency. But this does not mean that we need to rely on him as an indubitable authority. What I find appealing in a broadly Aristotelian account of abilities is that, as our aforementioned quotation illustrates, it posits a sort of definitional connection between the ascription of modal properties and the specification of the circumstances in which they are exercised.

Humean accounts of dispositions characterize this definitional connection with the help of counterfactual conditionals. In this, I think Humeans are right. But the Humean accounts also hold that the *ground* of that connection is not the possession of the relevant property itself but something else: some regularities, similarities between worlds or both. And in this, I wish to argue, the Humeans are wrong.

Identification, conditionals and reasoning

The first philosophers who attempted to provide a conditional analysis of disposition terms were the early logical empiricists. Importantly, their project was not to provide a semantic analysis of ordinary language but to explain how we can introduce theoretical-dispositional terms into the language of science. I suggest returning to this approach while abandoning the verificationist semantics which motivated the work of most early-twentieth-century empiricists.

In my proposed account, the purpose of conditional analysis is not to provide a semantic theory of the use of 'ordinary disposition terms'. Such terms – if it makes sense at all to distinguish them from any other terms in ordinary

---

[21] For further arguments which support the interpretation that Aristotle has a 'dispositional' – meaning, in this context, conditional – understanding of modal properties see Witt (2003, 42).

language that we use to ascribe properties to objects – are far too coarse-grained for our purposes. Since the sort of modal properties in which I am interested come in variable degrees of specificity, it is an essential task for any account of such properties to explain how we can *specify* them. And, as I already indicated earlier, the most important way in which we can specify them is to connect them to conditionals.

I shall argue that debates over the conditional analysis have been permeated with the mistaken assumption that if there is a conditional analysis of abilities (or dispositions), this entails the (logical) possibility of 'reduction' or 'elimination' of dispositions (or modal properties in general). I argue in Chapters 4 and 5 that this is a major misunderstanding. Even if all attempts for a conditional analysis failed, one could be a 'reductivist' or 'eliminativist' about dispositions, abilities or any other modal properties. In contrast, one can aim for an analysis without thereby committing oneself to 'reduction' or 'elimination'. In fact, I am going to show that a proper analysis of modal properties is not only compatible with, but also requires, realism about them.

All this does not mean that we should not take the objections to the traditional conditional analysis seriously. But the significance of those objections should be reassessed. This is the starting point of my arguments in Chapter 4. To give an initial hunch about the way in which, in my view, counterexamples to the conditional analysis should be treated, consider again the following (perhaps too) often discussed example of fragility. Suppose I say of a particular glass: it is such that it would break if it were struck. What if it is filled with protective material? The counterfactual is false, yet it seems that the glass is fragile. If a disposition or ability is, as is often said of such cases, 'masked', then the connection between the ascription of the relevant modal property and the truth of the conditional seems to be severed.

But suppose, we are not concerned with the meaning of the ordinary term 'fragility', but, rather, we would aim to identify an object's properties that ground its behaviour in various circumstances. Surely, the glass does *not* have the property specified by *this* counterfactual in *these* circumstances. But if I say the glass is now such that it would break if it were struck *and* were not filled with any protective material, then I specify a more generic property that the glass *does* have even in the circumstances in which it is filled with protective material.

This property is an ability that is more generic (less specific) than the ability identified by 'is such now that it would break, if it were struck' is in the sense that its possession is less sensitive to the variation of the actual (intrinsic or extrinsic) circumstances of the object. Thus, I shall argue, the problem of 'masks' is directly related to the issue of specificity: the *more* conditions we *explicitly* include in the antecedent of the conditional specifying an ability, the *less* specific (more generic) abilities we identify. By linking modal properties to certain conditionals, we can identify properties *of variable specificity* or variable degrees of determinateness.

Which ability – the more or the less specific – we aim to identify is governed by the context of our inquiry. Ontologically, the object may have both (logically non-independent, but nonetheless different) abilities, or may lack the more determinate/specific ability in the actual circumstances, while still possessing the more generic one. It is the practical or theoretical contexts in which we reason with possibilities that guide the selection of abilities that concern us. We might be interested in what can happen to this glass in the present circumstances; or we might be interested in what can happen to it more generally. We can ask: what would (have) happen(ed) if conditions c_1, c_2, etc. were (had been) satisfied? Or what conditions should we bring about in order for a possibility to be realized?

The significance of the variable genericness of abilities and the variable abstractness of possibilities will be explained in detail in Chapter 5. We shall see then the tight connection among the following three issues: (a) how abilities/liabilities of objects ground more or less abstract possibilities; (b) how the ascription of possibilities is linked to counterfactuals and thereby (c) how we can reason with possibilities (practically and theoretically). Since abilities ground possibilities, if there were no logical connection between the ascription of abilities and counterfactual conditionals, we could not explain how, by ascribing abilities, we can reason with possibilities.

As an example, suppose that I want to sweeten my coffee and I assume that this particular white solid cube in front of me is made of sugar. But in order to rationally entertain a course of action, I also need to know that if I put this piece of sugar into my coffee, it would dissolve, and that if it dissolved, it would sweeten my coffee. These conditionals connect the ascription of abilities of the sugar cube in specific context to the rationalization of my action. But they can do this only if the ascription of abilities (to the sugar cube, in this case) is somehow logically connected to some relevant conditionals that we use in our (in this case, practical) reasoning.

More generally, in the context of means–ends reasoning, it is possibility in the sense of contingency that we are interested in: what things can do here and now; and hence how we can use them in order to complete our ends. Such contingency is grounded in objects' dynamical-modal properties; that is, in properties that ground what they can do. And further, it is because the ascription of such modal properties entails the truth of certain counterfactuals that we can use our knowledge about objects' abilities in choosing our actions rationally in order to achieve our ends.

Earlier I mentioned 'masking' as one of the standard objections against the traditional conditional analysis. But I shall further argue that, generally, other standard objections should also be answered by applying a similar strategy: we should understand them from an ontological, rather than merely semantic, perspective. If we clarify what exactly happens in the world when cases which are mentioned as counterexamples to the analysis occur, we can easily reinterpret or reformulate it to avoid the objections.

Apart from masks, the two major standard objections to the conditional analysis derive from the observation that (i) abilities and dispositions might be, in contemporary jargon, 'finked'; and, more importantly, that (ii) abilities and dispositions might be – again, using the standard phraseology – 'mimicked'.

I shall argue that the first problem is related to the observation that things' modal properties might change at any moment, including the times when they are about to be exercised. For this reason, the occurrence of their exercise's result – which is, in such cases, an event – is conditional on their bearers' retaining the relevant property until the resulting event occurs. I shall explain, in several different contexts, the significance of this condition.

As far as 'mimicking' is concerned, I shall argue in Chapter 6 that examples of 'mimics', rather than being counterexamples to the conditional analysis, in fact support it by revealing something important about the nature of events and states which are the results of the exercise of modal properties in certain conditions.

Manifestations, events and actuality

Aristotle understands events as changes. He also seems to hold that what sort of changes happen in the world is determined by which capacities are exercised. I shall argue for a similar view about the connection between events and modal properties in Chapter 6. The nature of events, activities, interactions can be properly understood only if we take them essentially to be exercises or realizations of certain modal properties. This account of events is in sharp contrast with the currently most popular accounts, according to which particular events must be 'quality changes' or, alternatively, 'tokenings' of properties by an object at a time.[22]

---

[22] For the first, see Lombard (1986); for the second, see Kim (1993/1976). For a good summary of alternative metaphysical views of events that are relevant in this context see Mcdonald (2005, 181–215).

Changes of sensibly occurring qualities can, indeed, manifest, and hence help identify, events. But such changes do not capture the real nature of events since, as the phenomenon of 'mimics' proves, the same change of observable qualities can be the result of the exercise of distinct abilities. Sudden splintering is an observable quality change; but it can be either *breaking* or *exploding*, depending on whether the event manifests fragility or the ability to explode. Moreover, a lot of things can happen in the world without any occurrent 'quality change'. When things are in equilibrium, many abilities are exercised; but no 'quality change at a time' occurs.

Further, if we follow this account of events – inspired partly by Aristotelian considerations about movement and change – then we can explain how objects are 'involved' in events and also, what it means to 'token' a type of events. Objects (or substances) are 'involved' in events in the sense that, whenever an event happens, some object(s) exercise some abilities. And since events are typed with the help of the abilities which are exercised when they occur, 'tokening' just means exercising a modal property at a specific space and time. This means that we can understand the structure of reality, considered as a sequence of events, only if we can properly understand the underlying modal structure of the world.

The last point leads us to another objection to the modal property account of contingency, which I shall discuss in Chapter 7. Critics of such accounts often complain that the modal structure of the world logically presupposes an ontologically prior 'categorical', that is to say, nonmodal structure. Otherwise, such critics claim, the actual world would be a 'mere potentiality'. It is partly for this reason that the Humean account of contingency interprets possibilities as 'recombinations' of the allegedly fundamental, nonmodal, 'categorical' properties.

In the final chapter, I shall argue that this objection is based on several confusions. First, it is based on the unjustified assumption that we can draw an intelligible distinction between modal properties and 'categorical' ones. However, as mentioned earlier, I shall raise doubts about the very meaningfulness of categorical properties as fundamental properties in the world. In fact, it seems that all we know about the nature of those properties is that they are supposed to be nonmodal. We do not even seem to have any clear examples of them.

Sometimes shapes and structures, and other so-called primary qualities, are mentioned as examples of categorical properties.[23] But I shall argue that shape

---

[23] Primary qualities can of course be 'powers'. We owe to Mackie's interpretation of Locke the popular but mistaken view that the primary-secondary quality distinction corresponds to 'categorical-dispositional' one; see Mackie (1976), Chapter 1.

and structure are nonmodal only if we understand them as formal-mathematical representations. Spatiotemporally located shapes and structures are modal properties since their instantiations determine what things can and cannot do. A solid round object *cannot* fit into a squared hole. There is a sense in which structure and shape themselves, as well as some other physical properties, are not abilities. But they are such physical properties which are grounded in the more fundamental abilities of objects which instantiate them. The grounds of abstract physical properties like shapes and structures must certainly be modal, because their nature is determined by their contribution to the *possible* behaviour and interactions of material objects or particles.[24]

Hence, we cannot understand what 'actual' means with reference to the 'categorical'. The actual world consists in the spatiotemporally distributed stuff, and spatiotemporally located individuals constituted by it, which instantiate more or less specific modal properties. These properties determine the ways in which the world can change and develop, but neither chemical substances and particular objects nor their properties are 'mere potentialities'. Aristotle is certainly right that 'the actual is prior to the potential'. But what is actual is the distribution of concrete, maximally specific, spatiotemporal objects/substances in this world together with their modal properties (abilities and dispositions). It is this concrete actuality that determines which abstract possibilities exist in this word and in which ways the history of the world can develop. Consequently, contingencies are possibilities in this world, and they are not to be understood, as the world-account has it, with reference to other possible worlds identified by the distribution of merely categorical properties in them.

Contexts, method and ground

Although this book is written within the analytic tradition of philosophy, the way in which I approach some problems often differs from the one that is characteristic of many works in that tradition. While most typical analytic philosophy texts thoroughly engage with the current literature on their chosen

---

[24] Moreover, as already Leibniz acutely recognized, 'there are in fact no precise shapes [*certae figurae*] in the nature of things, and consequently no precise motions [*certi motus*].' He continues to argue from this important observation that the only genuine properties in space must be modal. 'And just as color and sound are phenomena, rather than true attributes of things containing a certain absolute nature without relation to us, so too are extension and motion. For it cannot really be said just which subject the motion is in. Consequently, *nothing* in motion *is real besides the force and power vested in things*.' Cited from Levey (2005, 83–4), my emphasis.

topic, they are rarely concerned with the context in which a philosophical problem originally arose. In my view, however, looking back on the history of those conceptual and theoretical changes which have led to our present concerns can help identify some of those presuppositions on which our current arguments, often implicitly, rely.

For this reason, I shall often start the discussion of some problem with a brief summary about its origin. It is a trivial, but nonetheless important, observation that all philosophical problems arose in a context. The concepts we use and the problems we try to answer have a history; their meaning derives to a large extent from the works and ideas of our intellectual ancestors. This is certainly true about the problems and concepts discussed in this work: the concepts of dispositions and abilities; and the problem of modality.

Where the relevant history begins can vary, however, according to the different issues. In some cases, it starts with Aristotle; in others, it starts with early moderns; most often, it starts with early-twentieth-century empiricism when the current philosophical concept of 'disposition' has been introduced. Central notions and ideas in this area, like the very concept of disposition, the meaning of 'manifestation', the idea of a 'conditional analysis' were introduced into philosophy in the early twentieth century for answering specific philosophical problems. However, since then, the philosophical context – in fact the whole philosophical atmosphere – in which these concepts and ideas are used changed significantly. Unless we are aware of these changes which may directly affect the philosophical content of the concepts we use, we can easily become confused about the meaning of the problems we must face.

The relevance of the origin of some concepts to philosophical inquiry has first been noticed by a philosopher who does not have a very high standing in the analytic tradition. It is something that I learned from reading, and trying to grapple with, Hegel's texts. That the origin and the context can be important for the deeper understanding of a philosophical problem is a matter of general outlook, as it were. But there is a more specific aspect of Hegel's way of thinking that influenced the ideas developed in this book too.

Abstractness and concreteness are often understood in philosophy as an all-or-nothing matter. Numbers and fictional characters are abstracts entities (if they are entities at all), while racehorses and pet dogs are concrete individuals. But abstractness and concreteness can also be a matter of degree. In Hegel's onto-logic the gradable distinctions between the *very* abstract and the *more* concrete, associated with degrees of determinateness, play a fundamental role. I shall also argue that metaphysics, and especially the metaphysics of modality, can benefit from interpreting abstractness and concreteness as gradable ontological characteristics.

A theory in metaphysics can start with what is the most abstract, and hence less determinate (for Hegel: the concepts of *Sein*, *Nichts*, *Werden*) and then can proceed towards the less abstract and more determinate (towards understanding the structure of *Dasein*) *via* concretization. This was Hegel's own grand – and, at least to my limited understanding, failed – project in his *The Science of Logic*.[25] But we can also follow an alternative path in metaphysics. We can begin our investigation with what is the most concrete and determinate: the properties instantiated by objects and substances located in space and time. It is then these concrete objects and their properties which ground, by their nature, the more abstract and indeterminate possibilities.

As already mentioned, this sort of 'progress' will play a crucial rule in my account of abilities and contingent possibilities. Abilities that ground possibilities can be more or less specific, depending on how many conditions of their exercise are satisfied in the concrete spatiotemporal world. If all are, then they are exercised, and concrete spatiotemporally located events occur in the world. If only one of them fails to obtain, the ability which the object has is maximally specific, and the possibility grounded by it is the least abstract unrealized possibility, since it is very close to the maximally determinate concrete reality. By specifying more conditions necessary for the exercise of an ability, we can ascribe more generic abilities which can ground more abstract possibilities. And by doing so, we move from the most concrete and determinate towards the more abstract and indeterminate.

Since according to the theory of contingency I explore in this book, abilities *ground* possibilities and dispositions *ground* behavioural tendencies, a few words seem to be in order to clarify the sense in which I use the notions of 'ground' and 'grounding'.

The idea of ground and grounding has recently become a topic of its own in meta-ontology. However, this book is not about meta-ontology – a study about the very meaning, relevance and methods of ontology – but a study in a 'first-order' metaphysical problem: the problem of modal properties and the nature of contingency. Thus, all I can, and aim to, do here is to justify my use of such 'grounding claims'. I do not try to engage with general debates over the nature and unity of the different sorts of grounding relations in metaphysical enquiry.

---

[25] I hasten to add: by saying that it was a failed project I do not mean that it is worthless or that it lacks many important insights. If judging a philosophical project to be a failure overall meant to hold it invaluable, we would have, I am afraid, little reason to read or do philosophy at all.

It appears to me that there are two questions that we need to distinguish about the use of grounding claims in metaphysical enquiry. The first is whether the notion has a legitimate application in metaphysical reasoning at all. The second is whether grounding is the ultimate explanatory relation in metaphysics either in the sense that every genuinely metaphysical explanation must have a unified form; or in the sense that they are all 'species' of one generic form of explanatory relation, which is grounding itself.

My answer to the first question is, as must be evident by now, an unqualified yes. The issue about the priority of certain sorts of facts or beings, as opposed to the ontological dependence of others, has always been a (perhaps *the*) central topic in metaphysics. To talk about grounds is just one useful way to express claims about such priority and dependence relations. There are other ways as well, of course. We can also express such claims about priority and dependence by saying that certain sorts of facts obtain *in virtue of* another; or that certain things are ontologically *more fundamental* than others.[26]

But whichever vocabulary we choose, serious metaphysics has always involved, in one way or another, the implicit thought that some kinds of beings or facts ground another kinds of beings or facts in the sense that the former are ontologically prior or more fundamental than the latter; irrespective of whether or not the use of these words or their cognates was in or out of fashion.[27] This seems to me to have been true even when other concepts, like 'property-identity' or 'supervenience', were more popular among analytic metaphysicians.

The idea of two (?) properties being 'identical' has always struck me as rather obscure. But even if we could make some clear sense of such claims about identity, they have rarely served the purposes of those who introduced them. For instance, philosophers who argued that mental states are identical with physical states did so in order to defend materialism about the mind. But then they must have been committed to the view that the 'identity' of mental and physical states entails the metaphysical priority of the material. Unfortunately, the notion of identity is inapt to express this kind of priority, for identity is a symmetrical relation (if it is a genuine relation at all). Hence, the implicit idea behind the

---

[26] 'More fundamental' needs to be distinguished from the question about what is fundamental *simpliciter*. The ground is in a sense more fundamental than what is grounded, but the notions of ground and ground*ing* can be used without the assumption that anything is fundamental *simpliciter*; just as one can explain things causally without assuming that there are 'ultimate causes'.

[27] Since the issue of using considerations about priority and dependence has become such a big issue only recently in analytic metaphysics – now that many have abandoned the Quinean dogma according to which the proper topic of metaphysics is 'existence' understood in terms of quantification – it might be worth mentioning that early analytic metaphysicians already used these notions without any scruple. See, for instance, Strawson (1959, 17).

'identity-theory of the mind' must have been that the instantiation of some material properties (e.g. some properties of the brain) grounds the instantiation of a person's mental properties.

Subsequently, the idea of identity has been superseded by supervenience, which is indeed a more useful concept for capturing the relation between different classes of properties. Supervenience, unlike identity, is a genuine relation which is compatible with the ontological difference of the types of properties related. Moreover, it is typically – even if not always – asymmetric. Hence, even if this was not explicitly expressed in this way, one of the intuitive appeals of the concept of supervenience in metaphysics seemed to be that its use provided a conceptual tool for discussing issues of priority and dependence.

Unfortunately, though introducing supervenience rather than identity was an advance in our understanding of certain metaphysical issues, it was soon recognized that the concept of supervenience, without explicitly filling it in, as it were, with some extra ideas about metaphysical priority, is far too weak. The merely formal notion of supervenience is easily applicable to characterize many contingent or even accidental correlations, which do not express any metaphysically important connection. Supervenience can be the result of nomic, rather than metaphysical, correlations. The temperature of gasses supervenes on their volume and pressure, but the latter does not 'ground' (in the relevant metaphysical sense) what their temperature is. Even worse, supervenience holds when correlations seem entirely accidental. Students' grades for courses supervene on the order of letters in the exam papers submitted. But a series of letters does not ground a grade in any sense (in fact, I doubt that there is any 'metaphysical ground' for grades at all).[28]

In contrast, whether wholes supervene on their parts, or whether the mental supervenes on the physical, are substantive and important metaphysical issues. Why? Because we are interested in a question about ontological priority and dependence. We aim to understand something about the ontic structure of reality. Metaphysical grounds must tell us something about the conditions of existence or about the nature of facts or things grounded; and mere supervenience cannot do this. For this reason, I do not think that I need any special justification for using the concept of ground and grounding. Serious metaphysics has always done that, even if only implicitly.

---

[28] This problem with the concept of supervenience in metaphysics was first made clear by Horgan (1993).

However, as I mentioned, there is another current debate over the notions of ground and grounding, which does not concern its meaningfulness and applicability in metaphysical enquiry, but the unity of the different contexts in which we apply it. More precisely, the question has been raised whether grounding is the ultimate explanatory relation in metaphysical explanations; or at least a 'genus' of such explanations in the sense that all specific explanations in ontology must be of its 'species'. This would mean that there must be some common logical and conceptual features which the various kinds of claims about what grounds what must share.

About this project, I am sceptical. The kinds of philosophical problems in which the question of metaphysical priority and dependence arises are far too multifarious to permit, or require, the use of a unified explanatory relation.[29] In some cases, we are interested in the question which facts ground some other facts, for instance, whether facts about the physical ground facts about the mental. In other cases, we are interested in how different ontological categories, like universals, tropes or persisting individuals are related; or how different kinds of individuals are related (whether 'simples' grounds 'complex wholes'; or rather 'wholes' determine the nature of their parts), and so on. Priority and dependence relations can be well expressed by saying that one kind of fact, property, individual or set, and so on grounds, and hence metaphysically explains, something else. But the exact nature of that connection can still be a further question and need not fall under one 'genus' or be of the same nature in every context.[30]

More specifically, this book argues for a theory of contingent possibilities according to which they are grounded in the fact that a kind of modal properties, abilities, are instantiated by certain substances, objects, persons or their ensembles in the concrete spatiotemporal manifold. Put otherwise, we can also express this claim by saying that truths about possibilities are grounded in truths about substances', objects', person' and their ensembles' abilities. Yet another way to express the same claim is that possibilities which are abstract and indeterminate are grounded in what is concrete and more determinate. I doubt that the meaning of these claims can be clearly grasped only if we force them into some Procrustean logic postulating a single universally applicable grounding relation.

---

[29] This does not mean that I would deny, as Jessica Wilson seems to do, that grounding is a genuine form of metaphysical explanation (see Wilson 2014). I only mean that it cannot be formally defined with reference to some more general characteristics.
[30] In this, I follow Koslicki (2015).

For this reason, I am going to use the notions of ontological grounding and dependence in the hope that their precise meaning and significance can be clearly understood from the context in which they are used.

Since the following chapters will discuss many different, even though interrelated, issues, it is perhaps helpful to end this one with a brief overview of how I shall proceed. In the next chapter, I will attempt to explain the motivation behind the distinction between the two types of modal properties in the study of which I am most interested: abilities which ground possibilities and dispositions which ground laws. I shall argue that the key feature of abilities as grounds of contingent possibilities is their variable specificity. And although there are various ways to specify abilities, I shall argue further that the most interesting sense of specificity can be cashed out only with reference to the conditions in which such properties would be exercised.

This interpretation of specificity then will lead us to two further questions, which I shall discuss, respectively, in Chapters 3 and Chapter 4. The first question is whether the actual occurrence of some circumstantial conditions can be constitutive of the instantiation of certain modal properties. I shall argue that there is a sense in which they are and that, for this reason, abilities and dispositions can be extrinsic. A more detailed analysis of the nature of extrinsic/intrinsic distinction in the context of modal properties will show that the distinction, contrary to the usual assumption, is not absolute but gradable.

If abilities can indeed be variably specific and extrinsic, there is a further question about how we can specify them. In Chapter 4, I shall argue that the traditional conditional analysis is still the best conceptual tool specifying the content of modal properties. That analysis exploits the conceptual connection between the ascriptions of modal properties on the one hand and the truth of certain counterfactual conditionals on the other. This connection has been challenged by various criticisms based on counterexamples. I shall defend the analysis against these objections and offer two new versions of it – one for abilities, and another for dispositions – which clarify the respective roles which these properties play in grounding contingent possibilities on the one hand, and laws of nature on the other.

Chapter 5 shall provide further reasons to use counterfactual conditionals in the specification of abilities and dispositions. It shows that the impossibility to identify all *conceivable* conditions that are necessary for the exercise of an ability does not undermine the significance of the analysis for the specification of modal properties *in a relevant context*. The chapter also argues that this account can explain away the examples of 'conditionless dispositions' and that it is apt to characterize the nature of nomologically indeterministic processes as well. Further, it also shows how certain logical peculiarities of subjunctive conditionals can be easily explained if we consider their truth to be grounded in objects' abilities.

Traditionally, it is assumed that modal properties are manifested by events which happen, or can happen, in the world. But how are such events related to the modal properties which they manifest? In Chapter 6, I shall argue that we need to distinguish two senses in which events manifest dispositions. In one sense, events are qualitative changes or non-changes in the world which indicate for us that some modal properties have been exercised. In another, and deeper, sense, however, events are essentially the exercises of some modal properties. It is for this reason that, in many cases, just by identifying which modal properties have been exercised when an event occurs, we can also explain it.

Finally, Chapter 7 returns to our fundamental question of why the modal-property account of contingent possibilities is superior to the standard possible-world analysis. I offer a hypothesis about the origin of the possible world analysis of contingency and argue that, to the extent reference to possible worlds can be useful for arguing about contingent possibilities at all, it must rest on some account of modal properties. The book ends with showing that the fundamentality of possibility-grounding modal properties (*i*) is compatible with the existence of some non-fundamental properties which are not modal; and that (*ii*) it does not contradict the idea that actuality is prior to possibility. For the fundamentality of modal *properties* is perfectly compatible, and in fact requires, with the fundamentality of other kinds of entities: spacetime, matter and the objects which are constituted of matter when certain modal properties are actually exercised.

2

Dispositions and abilities

> *It is the privilege of philosophy to choose such expressions from the language of ordinary life, which is made for the world of imaginary representations, as seem to approximate the determinations of the concept. There is no question of demonstrating for a word chosen from ordinary life that in ordinary life too the same concept is associated with that for which philosophy uses it, for ordinary life has no concepts, only representations of the imagination, and to recognize the concept in what is otherwise mere representation is philosophy itself. It must therefore suffice if representation, for those of its expressions that philosophy uses for its definitions, has only some rough approximation of their distinctive difference; it may also be the case that in these expressions one recognizes pictorial adumbrations which, as approximations, are close indeed to the corresponding concepts.*
>
> (Hegel, *The Science of Logic*, 12.130.)

In twentieth-century philosophy of science and philosophy of mind the most frequently discussed modal properties were objects' dispositions. In some respects, dispositions replaced the more traditional notion of causal powers. Convinced by Hume's criticism over the idea of 'necessary connections', many philosophers felt that defending causal powers would force them to join what they regarded as 'reactionary company'.[1] Dispositions might have seemed less reactionary because, as Hugh Mellor once wrote, they can be, like pregnant spinsters, 'ideally explained away, or entitled by a shotgun wedding to take the name of some decently real categorical properties'.[2]

More recently, however, the situation changed significantly. As a result of certain developments in philosophy of science and metaphysics, philosophers

---

[1] As Shoemaker, who is partly responsible for the resuscitation of interest in causal powers, once said in Shoemaker (1980, 109). A detailed early account of powers in philosophy of science which is critical to the Humean consensus is Harré and Maden (1970).
[2] See Mellor (1974, 157).

began to take seriously the idea that objects can have genuine modal properties. In philosophy of science, nomic properties figuring in laws have been interpreted as generic capacities, which need to be distinguished from objects' dispositions;[3] or potencies, which are essentially associated with nomic properties as ultimate grounds of fundamental laws.[4]

Abilities have always played a prominent role in the philosophical analyses of conscious human behaviour as well as in different accounts of free will. Some philosophers understood agents' abilities as a kind of dispositions. Others tried to explain dispositions in terms of objects' abilities.[5] Importantly, since on the current philosophical understanding, dispositions can be ascribed to inanimate objects just as they are ascribed to human agents, the latter project must allow that we ascribe abilities not only to human or animal agents, but also to inanimate objects.

In the present chapter, I am going to argue, first, that we need to distinguish abilities and dispositions as two different, though not entirely distinct, modal properties. Abilities are not a special kind of dispositions, and dispositions are not a special kind of abilities. Further, I shall argue that although there are semantic reasons to consider dispositions and abilities distinct, the main reason to distinguish them is metaphysical: they play different roles in ontology. Abilities are properties which ground contingent possibilities, while dispositions ground either nomic or habitual tendencies.

Introducing dispositions

The expression 'disposition term' as it is used in contemporary philosophy has been introduced by Rudolf Carnap.[6] Carnap and the logical empiricists were not interested in the metaphysics of modality; in fact, given their general attitude towards metaphysics as an intellectual enterprise, they were not interested in metaphysics at all. Carnap's original project was semantic and, to the extent that he was committed to a verificationist theory of meaning, epistemic.

---

[3] See especially Cartwright (1989, 1999).
[4] See Bird (2007b).
[5] For the former, see Ryle (1949), and more recently Smith (2003), Vihvelin (2004, 2013), Fara (2008); for the latter, see especially Vetter (2013).
[6] Not much earlier, C. D. Broad still discussed the problem of properties which we would call today 'dispositional' under the name of 'powers' or 'causal characteristics' of material substances. See Broad (1925, 430–40).

A large part of the language of science is not 'observational' in the sense that most predicates found in works of science are not used to report observations. Scientific theories employ statements which ascribe such properties to objects which cannot be directly observed. Typical examples include 'Salt is soluble in water', 'Copper is a better conductor than iron' and 'Iron becomes flexible when tempered at high temperature'.

We can observe that a piece of salt is dissolving in hot water. But we cannot observe salt's solubility at times when it is not dissolving. We can observe that a steel rod, after having been blended, quickly regains its original shape. But we cannot observe that it is flexible at times when it is not blended. Nonetheless, sentences like 'Salt is water-soluble' and 'Objects made of steel are flexible' make perfect sense and are often used in science. Carnap called predicates like 'soluble' and 'flexible' 'disposition terms'.[7]

Somewhat later Nelson Goodman recognized that if we are committed to the phenomenalist version of empiricism – as most early logical empiricists were – then not just terms like 'solubility' or 'flexibility', but *every* predicate that is meant to ascribe *enduring properties* to objects must be dispositional. For we can correctly describe objects by these predicates even at times when they are not actually observed.[8] But, from a phenomenalist perspective, an object's being red when it is not seen, or its having a certain length when it is not measured, is just as dispositional as its fragility, solubility, flammability, toxicity or soporificality.

According to phenomenalists, the content of disposition terms and enduring properties can be understood only with reference to those observation-events, like the occasionally observed episodes of breaking, dissolving, burning, poisoning or putting us asleep, which *manifest* them to us. And since objects' colour or length can remain as unmanifested as their solubility or flammability, on the phenomenalist account, the ascription of colours and spatial properties can make sense only if we can explain how the use of colour-terms and extension-terms is connected to episodes of observing colours or measuring lengths.

When Carnap and Goodman discuss the problem of disposition terms, they often use examples borrowed from ordinary language. But it is important to stress that their main concern was not to understand how disposition terms are used in *ordinary* language. Their project was to explain *how we can introduce* certain terms into the language of science. And, given their commitment to verificationism and phenomenalism, they held that disposition terms can

---

[7] See Carnap (1936, 440).
[8] See Goodman (1954, 40).

be introduced only through their logical connection to sentences that report observations.

However, not every philosopher in the first half of the twentieth century approached the issue of dispositions in this way. The analysis of dispositions played a central role in Gilbert Ryle's account of the mind, but Ryle was critical of phenomenalism,[9] and he was not interested at all in how certain terms are introduced in the language of science. His project was to understand mental concepts by comparing them to the use of dispositional predicates in ordinary language.

Consequently, Ryle's main concern was to understand how disposition terms are *actually used* in English, and not how they might be *introduced* into a language in order to serve some theoretical purposes. This is a major difference between the logical empiricists' and Ryle's approach to the problem of dispositions. And although Ryle's theory of the mind is hardly as popular as it was once, the way in which he approached the philosophical problem of dispositions has exercised a lasting effect on debates over modal properties.

It is a legacy of Ryle's philosophy that the problem of dispositions is often discussed as a problem about the meaning of *ordinary* disposition terms. Many accounts of dispositions are interested in the semantic analysis of 'ordinary language predicates' like fragility, solubility and flammability. It is important to note that these accounts still leave open the question *which* predicates are 'dispositional'. Is 'red' dispositional? Is 'knowing (that) p' or 'desiring (that) q' dispositional? It seems that no matter how we answer these questions, our answer cannot be based *merely* on semantic considerations about the use of 'red' or 'desire'.

More generally, if dispositions and modal properties are indeed metaphysically important, then the study of disposition terms as they are actually used in ordinary language cannot be sufficient for a philosophical account of modal properties; and especially, as we shall see later, for a philosophical account of abilities. Rather, like Carnap and Goodman, we need a theory about how such terms can be introduced into our language in order to answer some philosophical problem.

---

[9] David Armstrong classifies Ryle's view as a 'Phenomenalist and Operationalist' account of dispositions on the ground that Ryle denies that the ascription of a disposition entails that its bearer is in a particular state (see Armstrong (1968, 85–6). However, denying that dispositions are states of their bearers is one thing, being a phenomenalist about them is another. Ryle is quite explicit about this issue when he says that phenomenalism was 'in error from the start' (Ryle 1949, 236); and then he devotes the whole section of the chapter on sensation and observation to criticizing phenomenalism. Thus, there is no justification for interpreting him, as Armstrong does, as a phenomenalist. Indeed, Ryle denied that dispositions are states (and hence that their ascriptions entail the attribution of a property), but his theoretical motives were completely different from the phenomenalists. As we shall see, his reason for denying that dispositions are states was that he understood dispositions as quasi-nomological generalizations, which makes him more similar to some later functionalists than to any phenomenalists.

Yet, if we reject phenomenalism and verificationism, our strategy to introduce dispositions and abilities into our ontology must be somehow anchored in our ordinary practice of ascribing abilities and dispositions. In fact, as I shall argue in a moment, the relation between dispositions and abilities can, initially, be well clarified with the help of some observations about how we ascribe abilities and dispositions as properties to objects.

However, ultimately, we need to distinguish dispositions from abilities for a philosophical, and not only for some semantic, reason. The main reason for distinguishing them is that, though both dispositions and abilities are modal properties, they fulfil fundamentally different roles in our ontology. Dispositions, as I have mentioned, were often understood as replacing the more traditional notion of causal powers, which were associated with 'natural (or divine) necessities'.[10] I shall assign a different, though not unrelated, role to dispositions. But I also suggest distinguishing dispositions from abilities. For abilities, unlike dispositions, ground possibilities; and, in a sense, they are more fundamental than dispositions are.[11]

In the sequel, I shall argue that we ascribe an ability in order to identify what a person or an object *can* do; whereas we ascribe a disposition in order to express what objects or persons tend to do. As mentioned in Chapter 1, Aristotelian 'dunamis' understood as a statistical property is related to objects' dispositions; but Aristotelian 'dunamis' understood as the ground of contingency is not.

Being disposed to and being able to

If we reject verificationism, we must accept that the philosophical problem about modal properties has both a semantic and an ontological aspect. But the connection between the semantic analysis *versus* the ontology of abilities and dispositions is rather complex. For, if it is right to conceive abilities as grounds of

---

[10] Harré's and Madden's early work on causal powers bears the subtitle 'A Theory of Natural Necessity'. More recent realists about dispositions, like Ellis (1999) and Bird (2007), argue that laws of nature are metaphysically necessary because they are the consequences of properties that are powers. As mentioned before, Mumford and Anjum (2011), and later in more detail Anjum and Mumford (2018), argue that there is a special sense of 'dispositional modality', which is weaker than metaphysical necessity, but stronger than possibility. As it will become clear later in this chapter, I agree with their account in this regard, though I think dispositions ground laws rather than causation.

[11] Although he does not use the concept of power, Goodman seems to hold a similar view; see Goodman (1954, 50–4). Closer to the present chapter's view, Rom Harré says that 'to say that a thing has a power is to say what is possible for it, for that is what it is talk of its dispositions' (Harré 1970, 101). I shall challenge the second part of this claim, but fully grant the first. In this regard, my account is akin to Vetter (2013).

contingent possibilities, then we need to introduce into our ontology many more abilities than ordinary 'disposition terms' can express. Moreover, as we shall see, abilities are typically not expressed by ordinary 'disposition terms' at all. And even when some conventional terms can express both dispositions and abilities, they do so in very different contexts.

Consequently, we cannot ground our account of contingency merely on the investigation of how disposition terms are used in English. As we shall see, there are many more modal *properties* in the world than ordinary disposition terms. However, the limited usefulness of investigating the meaning of disposition *terms* in modal metaphysics does not mean that we cannot, at least initially, approach the problem of modal properties with the help of investigating how we can ascribe such properties by using some generalized disposition-ascribing and ability-ascribing locutions.

Perhaps the simplest and most common way to ascribe dispositions and abilities to objects is the use of locutions 'S is *disposed to* φ' and 'S is *able to* φ'. The differences in the use of these locutions provide an initial reason to distinguish abilities from dispositions. Of course, the differences in their use do not explain the difference between the two types of properties. In order to explain that difference, we need to identify their distinct roles in ontology.

According to a standard approach to the use of conventional disposition terms, ascribing a disposition is to say something about what objects having the disposition are disposed to do in certain circumstances. As David Lewis suggests – and as many subsequent analyses agree – in order to give a semantic analysis of ordinary disposition predicates like being fragile or being toxic, we first need to make their meaning more precise with the help of a paraphrase. For instance, 'Arsenic is toxic' should be paraphrased as 'Arsenic *is disposed* to poison those who ingest it' (Lewis 1997, 153).

I suggest that the Lewis-type paraphrases provide the most generic forms in which we can ascribe modal properties to objects (or to substances, sets of objects, etc.). For we can apply them even when we do not have a specific 'disposition term' at our disposal to express a modal property.

Some philosophers objected that the 'is disposed to' locution is too technical for representing adequately the content of ordinary disposition terms like 'fragility' or 'toxic'.[12] This might be right, but as we have seen, the very notion of dispositions as used by philosophers is a term of art. According to its primary

---

[12] See Vetter (2015, 65).

meaning, dispositions were only used to describe persons' moods and character.[13] But if 'being toxic' *is* interpreted as a dispositional predicate in the more technical sense in which philosophers are interested, then there is no reason to deny that it can also be expressed by the paraphrase 'being disposed to poison'.

I shall use '... is disposed to ...' as the generalized form of disposition ascription, because with its help we can ascribe dispositions that cannot be simply expressed by ordinary disposition terms. An unripe lemon and an unripe apple have different dispositions: one of them *is disposed* to become yellow while the other one *is disposed* to become red.[14] But there are no ordinary disposition terms expressing the disposition to turn yellow and disposition to turn red.

In fact, it is hard to imagine how we *could* distinguish the modal properties we often need to distinguish without employing the '... is disposed to ...' locution. I shall use then the paraphrases of conventional disposition terms in forms of Lewis's explicitly dispositional locutions[15] as the most generic template for ascribing dispositional properties. An explicitly dispositional locution specifies the meaning of a disposition D. An object has D if and only if it is disposed to φ in some circumstances C.[16]

In what follows I shall also use what might be called 'explicit ability locutions'. By this I only mean that, just as the most generic way to ascribe dispositions is '... is disposed to φ ...', the most generic way to ascribe abilities is '... is able to φ ...'. Ascribing abilities in this way may sound initially somewhat artificial. As I have mentioned, abilities are most often ascribed to human agents. And even if we do attribute abilities to animals, we rarely ascribe them explicitly to inanimate objects. However, as we have seen, the same was once true of dispositions. Moreover, when we understand how abilities as properties are related to dispositions, ascribing abilities to inanimate objects will sound more natural.

In ordinary English it is entirely natural to talk about what things *can* do. Aspirin can release us from fever and headache, a train can take us faster to our destination than a coach does and a hot stove can melt ice. What I suggest is

---

[13] In fact, Ryle's idea was to understand such traits – and mental properties in general – by comparing them to terms such as 'fragility' or 'solubility'. It is partly as a result of this comparison that philosophers use 'disposition terms' to describe modal properties of objects, some of which would have been called 'powers' or perhaps 'capacities' before talks about powers became unpopular.

[14] Interestingly, these are Vetter's own examples, see Vetter (2015, 94).

[15] Following Choi (2008).

[16] Lewis, and many other philosophers following him, include as an essential part of the explicitly dispositional locution the phrase 'in response to stimulus σ'. I omit this part for reasons that will become clear later in Chapter 5. As we shall see, reference to 'stimulus' is essential only if we assume that dispositions must be *causal*; and even then, as far as the specification of generic modal properties is concerned, the selection of one condition as 'stimulus' is context relative.

that, in a technical-philosophical sense, we are permitted to express the same claims by employing the phrase '... is able to ...'. And further, if by using '... is disposed to ...' we ascribe a modal property which is a disposition, then by using '... is able to ...' we can ascribe a modal property which is an ability; no matter whether we ascribe them to animated or inanimate objects. In fact, as we shall see, what justifies the technical extension of the use of this locution is precisely the connection between 'can'-claims and the attribution of modal properties.

Using then such explicit disposition-locutions and ability-locutions, first, I shall argue that statements ascribing dispositions have an intuitively different sense from statements ascribing the 'corresponding' abilities. Subsequently, I shall argue that this difference is a consequence of the fact that having an ability to φ in C is not the same property as being disposed to display behaviour of type φ in the circumstances C. Tautologically, if an object or a person is disposed to φ, she or it has the disposition to φ. But having an ability to φ does not invariably imply that something or someone is disposed to φ. There is an important difference between the ascription of abilities and claims about what things are disposed to do.

Why abilities are not dispositions

Though I suggest that it makes good philosophical sense to ascribe abilities to inanimate objects as well, I shall first illustrate the difference between abilities and dispositions with an example involving human agency. Subsequently, I shall argue that a similar distinction should also be drawn in the context of inanimate objects, (chemical) substances and ensembles of objects.

Imagine Amy and Bob sitting at their dining table. As usual, they have salt and pepper on their table next to a bowl of soup, the first course of their dinner. Amy dislikes food which is not sufficiently salty as much as she dislikes spicy food. Bob must be cautious with salt – he has problems with his blood pressure – but loves spicy food. The following seems to be intuitively true of Amy and Bob in this situation:

- Amy is able to flavour her soup with pepper, but she is not disposed to.
- Bob is able to flavour his soup with salt, but he is not disposed to.

Moreover, since Amy does (intentionally) add salt to her soup, she must be able to flavour her soup with salt. And since Bob does (intentionally) add pepper to his soup, he must be able to flavour his soup with pepper. Intuitively, they have exactly the same abilities: both are able to flavour their food either with pepper, or with salt, or with both, or with neither. But they are not at all disposed to act in the same way. Amy is not disposed to pepper her food, while Bob is not disposed to salt it.

What explains this difference? Ryle – who otherwise interpreted every modal property as objects' or persons' dispositions – mentions that some sentences, which do not describe states or processes, ascribe abilities, capacities or liabilities to certain objects or to certain kinds of objects; while some other express behavioural tendencies, habits or proneness to do certain kind of things (Ryle 1949, 131). Unfortunately, Ryle immediately spoiled the importance of his own observation by narrowing down his investigations to the use of 'disposition terms', which did not allow for him to distinguish different types of modal properties and resulted in some confusion about their nature.[17]

Dispositions are indeed, as Ryle says, essentially connected to behavioural tendencies. To say that an object is disposed to do something is to claim something about how it tends to behave. Dispositions are modal properties because the tendencies need not be *actually* displayed in order to ascribe them to an object. An object is disposed to φ, and hence has the disposition to φ even when it is not φ-ing; or, as we shall see, even if it never actually φs. Thus, although in a sense explained later, dispositions are statistical properties, dispositions are modal properties as well.

In some cases – in the case of habitual dispositions – the ascription of a disposition requires actual frequency of some behaviour. In other cases, tendencies are interpreted counterfactually. A fragile object cannot break regularly; it can break only once. But *fragile objects* do have the tendency to break; and an object is fragile, that is to say, it is disposed to break, only if it is made of a kind of material which (in a certain state) does have the tendency to break in a relatively wide range of circumstances. In contrast, although a sturdy, that is, non-fragile, object is *able* to break in certain *specific* (and statistically atypical) circumstances, it does not have the tendency, actual or counterfactual, to break.

It follows – as I shall explain in more detail in Chapter 4 – that I interpret tendency as a statistical notion which I propose to cash out in terms of frequencies. I should emphasize that interpreting the concept in this way is partly a matter of theoretical decision and it is not based exclusively on some

---

[17] For Ryle, of course, this might not be a confusion. His project was to give an analysis of all modal property ascriptions in terms of hypotheticals, and the first step in the analysis was to interpret all such ascriptions as 'dispositional'.

observation about how we use the word 'tendency' in everyday language. As I shall argue later, I prefer the statistical interpretation because it allows us to give a relatively precise account of the conditions of disposition ascriptions.

Nonetheless, I do not think that the statistical notion of tendency is incompatible with our ordinary practice of ascribing dispositions. At least, I think that the possible counterexamples to it are somewhat dubious. It might be objected to my interpretation of tendency that it is incompatible with the fact that contraceptive pills have the dispositions to produce thrombosis even if this happens in about one in four thousand users.[18] Since dispositions entail tendencies, taking contraceptives must also have the tendency to produce thrombosis even if such cases occur very infrequently.

However, I do not find it natural at all to say that contraceptive pills are *disposed to* produce thrombosis. In fact, if this were true, we should never allow to use them. Contraceptives *can* produce thrombosis – they have the (li)ability – but it would be a mistake to claim that they are disposed to. Compare: would it be natural to say that peanuts are *disposed* to kill humans just because eating a small quantity of peanuts *can* kill some people? There might be a tendency to produce thrombosis in case of women in some specific and yet unknown physiological condition, but that is a different issue.

In fact, this case supports rather than challenges my distinction between abilities and dispositions. When we read information about possible side effects of medications, they tell us that they *can* produce this or that adverse effect in one out of 100, 1,000 and so on cases. It is never said, and for good reason, that they *tend* to cause such effects. If they had the tendency, they should never be used. In order to be aware of some risk or danger, in many contexts, all we need to know is what certain medications *can*, or are (li)able to, do in certain conditions. If this always entailed that they are also disposed to, and hence have the tendency to, harm us, we should simply avoid using them (unless we are deliberate risk-seekers, but that is a different issue.)

Thus, in the case of inanimate objects just like in the case of animated ones, we must distinguish objects' abilities from their dispositions. A sturdy object is not disposed to break, even if it is able to break. The ascription of abilities is naturally associated with what objects *can* do; the ascription of dispositions is associated with what they tend to do.[19] It follows that being able to φ does not

---

[18] I thank for this objection to one of my referees.
[19] In what follows, for the sake of simplicity, I shall often talk only about *objects'* modal properties. But as we shall see in the next chapter, chemicals substances, volumes of matter, animals, persons and systems of objects are also bearers of abilities and dispositions.

entail that something is disposed to φ. This is so because abilities ground what is possible, but a possibility does not entail any tendency.

There are several further important differences which follow from this fundamental one. First, we need to distinguish between these two facts: an agent's *not being able to* do something and her *being able not to* do something. The latter means that she is able to avoid doing certain things and hence describes a certain possibility; the former, in contrast, means to express a kind of impossibility. Similarly, it is not true of a hard and sturdy object that it is *not able to* break; but, if an object is sturdy, it must be *able not to* break in various circumstances in which fragile objects would break; that's why the object is sturdy, after all.

In contrast, if an object is sturdy and therefore it is *not disposed to* break, then it is also *disposed to not* break. The reason why 'not being able to φ' is distinct from 'being able to not φ', while 'not being disposed to φ' is not distinct from 'being disposed to not φ', is that ability-ascriptions and disposition-ascriptions have different roles in our language. Abilities are ascribed because they ground certain possibilities, and being possible that not-p is different from not being possible that p. The latter identifies an impossibility (the impossibility of p), the former a possibility (of not-p). In contrast, dispositions ground behaviour tendencies and not mere possibilities. And there is no difference between a tendency of not displaying a certain kind of behaviour and not having the tendency to display it.

This difference also explains why the ascription of an ability does not entail the ascription of the corresponding disposition. That something, like a steel rod, is able to break does not entail that it is disposed to break. For something can break even if it does not have the tendency to break. But an object can have a disposition to φ only if it has a certain tendency to φ.

Since dispositions ground tendencies, some dispositions can come in degrees. Some objects are more fragile, and hence less sturdy, than others; some objects are disposed to be scratched more easily than others; and some people are stronger smokers, that is, they have a stronger tendency to smoke in various circumstances, than others. Tendencies are matters of degree. Possibilities either exist or do not.

Thus, abilities are an all-or-nothing matter. Something can break if it has the ability to break. This is not a matter of degree. It will play a crucial role in my account of contingency that abilities can be more or less specific. Someone might have the ability to swim, the ability to butterfly, the ability to butterfly fifty meters within one minute and so on. Obviously, having the more generic ability does not entail having a more specific one. But abilities *themselves* do not come in degrees: someone or something either does have a (more or less generic) ability or does not.

The following objection has been raised against my claim that abilities do not come in degrees.[20] Sometimes we seem to compare people in respect of their abilities. We use comparatives like 'She can, that is, she is able to, do this or that better than me'. This seems to imply that abilities can come in degrees after all. I think, however, that such comparisons should not be understood as qualifying the abilities themselves. We use such comparatives to evaluate persons by comparing of what they *can* do. However, the ground of this evaluation is that 'She can – is able to – φ in ways W1' and 'I can – am able to – φ in ways W2'. The 'can' or 'able to' in this case is absolute. Different people can have different abilities because the levels or the qualities of achievement which their abilities' exercise represents are different. But they are either able to do something with a certain quality, at a given level, with a certain speed etc. or not.

Do more specific abilities always entail the less specific ones? It might seem that we have counterexamples to this entailment. For even if someone is able to swim in the Dead Sea, she may still not have the seemingly more generic ability to swim.[21] If someone is able to fly with an airplane, it does not seem to entail the more generic ability to fly.

In my view, such counterexamples are based on an ambiguity about some words of action. They do not directly concern the metaphysical issue I am concerned with here. If we disambiguate the meaning of these sentences, we can mean either that the kind of activity that we express as 'Dead-Sea-swimming' is not swimming at all, but rather a sort of floating; or that the agent can swim after all just as a sturdy object can break. The choice depends on what we mean by 'swimming'. The same is true about flying. We can mean that flying by an airplane is not flying in the sense in which birds fly; or we can say that it is, but then it is true that pilots (or airplane passengers) can fly.

But why does an ability to φ in C not entail the corresponding disposition to φ in C? Certainly, a tempered steel rod is able to break, but only if it is under very special, 'abnormal' or 'extreme' circumstances; for instance, if it is cooled down near to the absolute zero. However, in those circumstances, it is also disposed to break. Thus, if we specify the relevant circumstances C in which o is able to φ, then it also seems to be true that o is disposed to φ in C.

---

[20] By a referee of this book.
[21] Thanks for this example to Simon Rippon.

However, while this might be so in certain cases, it is not true in every case. It is obviously false about human agents' abilities. There are many things I am perfectly able to do now – smoke tobacco, have a shot in the morning, insult a stranger on the phone – which I am not at all disposed to do.

Some philosophers suggested that one can have the ability to φ in C only if one also has the disposition to φ when one tries to φ in C and that someone's ability to do certain things should be understood as being logically equivalent with this more complex, conditional dispositions.[22] But, in fact, an ability is not such a 'conditional disposition'.

Importantly, a conditional disposition is not a disposition analysed conditionally, but a disposition that one would acquire in certain counterfactual conditions. Consider someone who is a non-smoker: she is not disposed to smoke in circumstances in which a smoker would be disposed to smoke. But she might still be able to smoke in those circumstances. If conditional dispositions were indeed abilities, this should mean that she *becomes disposed* to smoke in those circumstances in which she tries; which would mean that by *trying* to smoke she *acquires* a disposition which she lacks otherwise.

But this could not be right. How dispositions are acquired is a moot question, but certainly no one becomes a smoker just by trying to smoke once, even if one succeeds in doing so. More generally, a person's ability to φ does not imply that she would acquire certain dispositions in the circumstances in which she would φ. I am able to type the word 'synesthesia'. But this does not mean that if I tried, then I would become *disposed* to type it. What might be true, if anything, is only that, in the proper circumstances, I *would succeed* typing it, if I tried.

Such examples concern specifically human abilities, and it might be said that, at least in the realm of inanimate objects, having an ability must entail the corresponding disposition. But, although the difference between the ascription of dispositions and the ascription of abilities is perhaps most conspicuous in the realm of intentional human actions, it applies equally in many other contexts.

A dog is able to breastfeed a young lion, but it is not disposed to; cacti are able to grow in my study, but cacti are not disposed to grow in my study. Peanuts are not toxic. They are not disposed to poison people. In fact, they are tasty and nutritious. That is why many meals served in restaurants contain peanuts. But peanuts can poison people; people who are allergic to them. That is why it needs

---

[22] See Fara (2008). Fara is interested in the reinterpretation of the conditional analysis of free will, and not in the problems of abilities in general. He claims that his proposal gives a good response to some objections against the traditional conditional analysis. But what he says can be extended to abilities more generally considered too.

to be indicated which meals contain peanuts: for, although they are not disposed to, they *can* harm people.

Gold is not disposed to dissolve in acids. But gold can be dissolved if the acid is *aqua regia*. For *aqua regia* does have the ability to dissolve gold. In fact, gold is disposed to be dissolved by *aqua regia*. But from the fact that gold is disposed to dissolve in *aqua regia* it follows only that gold is able to, or *can*, be dissolved in acids; it does not follow that it is disposed to be dissolved in acids. Gold's traditional value has been derived partly from its disposition to resist certain chemical reactions to which other base metals are exposed. Acid tests have been used for centuries to prove that something is (made of) gold precisely because gold is not disposed to be dissolved in acids. But this does not mean that it *cannot*, that is, not able, to dissolve in (some) acid.

The reason why we might be prone to believe that the ability to φ entails some corresponding disposition to φ, especially in the case of inanimate objects, seems to be epistemic. The most common way to learn about physical objects' and chemical substances' dispositions is, after all, to observe how they regularly behave in certain circumstances. Gold is often observed to dissolve in *aqua regia*. And it seems right to say that it is disposed to dissolve in *aqua regia*. It is for this reason that we are *justified* to think that it is also able to dissolve in *aqua regia*. But even if *aqua regia* is composed of acids, this does not mean that gold is disposed to dissolve in acids *simpliciter*.

This is not to deny that we can distinguish more and less specific dispositions, just as we can distinguish more and less specific abilities. But a crucial difference remains. If being disposed to φ in C were to specify an object's ability, then the ascriptions of specific dispositions should entail the ascription of the more generic one(s); since, as I have just argued, the ascription of more specific abilities does entail the ascription of more generic abilities.[23] However, it can be true of Bob that he is disposed to swim when he accidentally falls in deep water, but false nevertheless that he is disposed to swim. His accidental swimming entails only that he is able to swim, not that he is disposed to.

More generally, if an object has the ability to φ in C then it must also have the ability to φ because the possession of more specific abilities entails the less specific ones. But since tendencies in some specific circumstances to φ do not logically entail the tendency to φ *simpliciter*, an object's dispositions to φ in C

---

[23] I shall say more on the significance of this implication for an account of contingent possibilities in the next chapter.

does not entail that it is disposed to φ *simpliciter*. They only entail that the object *can* φ.

Further, and relatedly, most objects which have a certain disposition can have what has recently been called the object's 'Achilles heel' with respect to that disposition.[24] It seems that even very fragile objects can fail to break in circumstances in which they usually do. This means precisely that they can, that is to say, they are able to, not-φ, even though they are disposed to φ; and they are able to not-φ even if they are not at all disposed to not-φ. This is so because having a tendency to φ in C need not entail the inability to not φ in C.

What all this shows is that we cannot simply 'pair', as it were, the ability to φ in C with the disposition to φ in C. If an object is able to φ but its ability would be exercised only in exceptional circumstances, it is misleading to claim that it is disposed to φ *simpliciter*; but it might be true of the object that it *can* φ precisely because it is able to φ in those circumstances. Most medicaments can cause harm to some patients in certain circumstances; for instance, when they are overdosed or when they are taken by someone who is allergic to them. But medicaments are not 'disposed to harm people as a response to being ingested'; their ability to harm does not make them toxic. In contrast, even if some toxic material can cure people in certain special circumstances, they are not 'disposed to heal'.

The difference between abilities and dispositions can also be illustrated by the ways in which we can specify them. As mentioned, the ascription of more specific abilities entails the ascription of less specific ones. If someone is able to speak loudly in a noisy restaurant, she is able to talk loudly; and then she is able to talk. And if something is able to dissolve in *aqua regia*, given that *aqua regia* is an acid, it is also able to dissolve in acids. But this is not true of dispositions. As we have also seen, if something is disposed to dissolve in *aqua regia*, this does not entail that it is also disposed to dissolve in acids; and that someone is disposed to talk loudly in a noisy restaurant does not entail that she is disposed to talk loudly; or disposed to talk at all in the sense in which this implies that she is generally talkative.

Of course, there is much more to be said about the precise nature, significance and implications of the degrees of specificity of modal properties. This is the job for the next two chapters. To conclude this section, I only want to indicate an important difference in the ways in which some human abilities and in which some dispositions are acquired.

---

[24] The term has been introduced by Manley and Wasserman (2008). I will return to the significance of Achilles heels in the context of conditional analysis of modal properties later in Chapter 4.

It is interesting to observe that while many human abilities are learned, dispositions are typically not acquired by learning. I have once learned to swim, but I never 'learned' to swim regularly. In order to learn to swim, I had to have the prior (unlearned) ability to be able to learn to swim; but while I can be disposed to swim only if I am able to swim, the acquisition of that disposition does not require a prior disposition to learn to swim regularly. More generally, it might be necessary for the acquisition of certain abilities that someone have the ability to acquire them. But it does not seem to be necessary for the acquisition of dispositions that someone have a prior disposition to acquire them.

Why dispositions are not abilities

So far, I have argued that having an ability to φ does not entail having a disposition to φ. Thus, abilities cannot be understood as a specific sort of dispositions. Nonetheless, we might try to understand dispositions as some kind of abilities. *Prima facie*, there appears to be a logical or conceptual connection between the ascription of abilities and behavioural tendencies. For the tendency to φ seems to entail some ability to φ.

But does indeed an object's or a person's disposition to φ entail that it or she is able to φ? Ann can be disposed to swim (regularly) even when she is not able to swim because her leg is broken or because there is no suitable swimming pool in the vicinity. And Bob can be disposed to smoke even at times when he does not happen to have tobacco; or when he sits in a place where smoking is not permitted. One can be disposed to φ even in such circumstances in which one is not able to φ.

One can be disposed to φ at time t even if she is not able to φ at t because abilities ground possibilities which can easily change if their bearers' states or circumstances change. In contrast, dispositions are statistical properties, the possession of which is less sensitive to the alteration of their bearer's circumstances. A tendency to φ is not undermined by an object's or a person's inability to φ in some specific conditions. It is for this reason that they can retain a disposition to φ even if, in some specific circumstances, they are not able to φ.

Nonetheless, one might argue that dispositions are *a special sort* of abilities. Consider the case of fragility. As mentioned, almost any solid object is able to break in *some* circumstances. However, it seems that certain, or certain kind of, objects break *more easily* than others. A steel rod or a piece of marble can break, but there are many more circumstances in which a China vase can break.

Dispositions might then be interpreted as 'easy abilities'.[25] If the ability to break comes in degrees, then fragility is the ability to break easily; or rather, the ability to break more easily than something else does. It is easier to break a China vase than it is to break a tumbler, and it is easier to break a tumbler than it is to break a (good) chair. The easier it is to break an object, the more fragile it is.

There is certainly something important about the observation that some dispositions can come in degrees. If an object's disposition is the property that grounds its tendency to display a certain form of behaviour, then dispositions are essentially statistical properties. And statistical properties can come in degrees, since objects can tend more or less to display certain forms of behaviour. But this does not mean that a disposition is a kind of ability.

First of all, it is not at all clear that 'easily' or 'more easily' can be applied to every disposition. Its applicability depends on whether or not a disposition comes in degrees. Many macro-physical dispositions (like fragility) and chemical dispositions (like solubility) do come in degrees. But nothing can be a predator or a ruminant to certain extent. A wolf cannot hunt 'more easily' than a cow does, and a cow cannot ruminate 'more easily' than a wolf does.

Much more importantly, as I have argued earlier, if having a possibility is not a matter of degree, then it is hard to interpret the relevant sense of 'easy ability'. Suppose an object can break easily or can break more easily than some other. This does not mean that the *possibility* that it breaks is easy. An object can break in many different circumstances. Thus, we might say that an object can *break easily* if it would break even if only a relatively weak force were exercised on it. But this does not seem to qualify the *possibility* itself. For it is hard to understand how a possibility can be so qualified. A possibility either exists or does not. It cannot exist 'hardly' or 'easily'.

When we say that something 'breaks easily' what we mean is that it is disposed to break more easily than some other objects; in other words, it has a stronger tendency to break. 'Easily' can modify a disposition. It is also true that if something has the disposition to break easily then it can break easily in the sense that it has the ability to break in many such circumstances in which a harder object is not able to break. However, 'easily' does not modify possibilities or the corresponding abilities. Whether or not an object is able to break – or is able to break in this or that circumstances – is an all-or-nothing matter.

English simple present – as opposed to the progressive – often expresses habitual dispositions (Bob smokes; Ann swims, my kids go to school, cows

---

[25] As Barbara Vetter argued in Vetter (2014, 2015).

ruminate, etc.).²⁶ Similarly, when we say that salt *dissolves* in water, we mean that it is disposed to dissolve in water, and we can easily dissolve it in water. But it does not seem to be generally true that whenever the simple present entails a disposition, then it also entails 'can easily'. When we say 'Birds lay eggs' what we mean is that (female) birds are disposed to lay eggs (or that they are oviparous); but we do not mean that they can easily lay eggs. Or when we say that Jim lifts heavy weights (he is disposed to lift them given that he is a weightlifter) we don't mean that he can easily lift heavy weights. In general, the use of simple present often entails ascription of dispositions without entailing 'easy possibilities'.

However, my point is not really about language, or, specifically, how 'easily' and 'hardly' are, or can be, used in English. My point is about the metaphysics of properties. The idea that motivates the attempt to understand dispositions in terms of abilities is that the latter ground 'restricted possibilities'. And there is indeed a sense in which some possibilities might be 'restricted'. But the restriction is not a matter of the *degree* of possibilities but the matter of their *abstractness*. With 'restrictions' we can identify more or less abstract possibilities. And possibilities at a certain level of abstraction are neither easy nor hard. They either exist or do not.

We might classify *terms* as 'dispositional' and 'non-dispositional'. It is not entirely obvious, of course, that terms can indeed be so classified, since, as we have already seen, which terms are assumed to express a disposition might depend on one's prior metaphysical commitments. For some, 'red' is non-dispositional; for others, it is obviously dispositional.

Suppose though that this classification does make sense. Then it is true that 'disposition terms' can express abilities as well. In English, disposition terms might be formed virtually from any transitive verbs with the help of applying affixes like '-able' or '-ible' (perceptible, drinkable, edible, soluble, flammable, etc.).²⁷ But the classification of such terms as 'dispositional' should not obfuscate the important ontological difference between abilities and dispositions as modal properties. Whether or not we use a term to express abilities or dispositions is determined by the context, and I suggest that the difference between the two can

---

[26] About the philosophical significance of the difference between simple and progressive tenses in distinguishing processes and dispositions, see Thompson (2008), Chapter 8, esp. 122–8.

[27] As has been already observed by Quine (1960, 224); see also Vetter (2014, 2015). The same is true, for instance, about Hungarian which also has a special modifier with which verbs can be turned into 'disposition terms'. We need to add, of course, that many disposition terms (e.g. 'fragile', 'migrant' or 'nice') are not so formed and that we often make claims about dispositions and abilities without using such terms.

be made explicit by applying the more precise locutions: '... is disposed to ...' and '... is able to ...'.

The same is true about the use of 'can'. 'Can φ' typically expresses the ability to φ, but in certain contexts it is used to express a disposition. There are contexts in which 'can easily' is indeed used to ascribe a disposition: something can break easily (that is, it is fragile) or someone can get angry easily (that is, she is irascible). I am less certain whether it makes sense to say that salty water can *easily* conduct electricity; or birds can easily lay eggs or hunt; or that red objects can easily elicit red sensations. An honest person is disposed to tell the truth. Perhaps she can sometimes easily tell the truth. But telling the truth is sometimes difficult – that is to say: not easy – even for an honest person.

In our earlier example, Amy has the disposition to salt her soup and Bob has the disposition to spice his. And they do have those dispositions even when there is no salt or pepper on their table and even when they have no access in general to salt and spice. In such circumstances they are certainly not able to salt or spice their food. Yet, they can retain their dispositions to do so. It is hard, however, to understand what it would mean to claim that nonetheless they have their respective dispositions because they can 'easily' salt or pepper their food. It seems that in the relevant circumstances they simply lack the ability to do so.

We can call Amy's and Bob's dispositions *agent-relative* dispositions. They have their respective dispositions because they have *their own* reasons to do or to avoid certain things, and having those reasons, they are also, *ceteris paribus*, psychologically disposed to do or to avoid doing them. But many dispositions are not agent-relative, but nomic. And my main concerns in this book shall be such dispositions because they have the role to ground the special, nomological modality of the occurrence of certain events. However, as we shall see, nomic dispositions are often relative to kinds. Abilities, in contrast, are not relative to a kind.

Human beings as belonging to a biological kind with certain type of physiology are disposed to digest cow milk. This is why cow milk, unlike grass and petrol, is considered to be nutritious. However, there are some people who cannot digest cow milk. Do they thereby lack the disposition to digest it? If they did, they would not be disposed to digest cow milk just as they are not disposed to digest grass or petrol. But obviously, there *is* a difference here. As humans, they must be disposed to digest cow milk even if, in the more specific conditions in which they are, they are not able to.

Now, what are those more specific conditions? They are the conditions which make them lactose-intolerant: not having, or not having sufficient amount of, lactase enzyme in their body. Being lactose-intolerant does not deprive humans of their generic disposition to digest cow milk as opposed to grass or petrol. It only shows that, given their specific conditions, they cannot or not able to. Hence, the ascription of a disposition does not entail the ascription of the corresponding ability.

However, we can alter the relevant context in which we ascribe a disposition. Lactose-intolerance is a *medical* kind. When we say that lactose-intolerant people are not disposed to digest cow milk, we consider them as members of a subgroup of humans. They belong to the medical kind 'lactose-intolerant humans'. And as a member of that more specific subgroup, they are of course not disposed to digest cow milk or dairy products in general. This feature of kind dependence of dispositions will play a crucial role in my proposal about how we can identify nomic dispositions through their connection to counterfactual conditionals.

Nonetheless, it seems that nothing can have the tendency to φ if it does not have at least the *generic capacity* to φ. No one who lacks the 'generic capacity' to swim can be a swimmer. And nothing that is 'generally incapable' of dissolving can be disposed to dissolve in acids. And are such capacities not just very generic abilities? Perhaps all I have shown is that the connection between abilities and dispositions must obtain at the appropriate level of generality, not that the ascription of dispositions does not entail that an object must possess the relevant abilities. Even if Bob need not have the specific ability – 'the ability to smoke in C at t' – in order to be a smoker (and hence in order to be disposed to smoke), he does need to have the most general ability at t to smoke (at some time or other); otherwise, he could hardly be a smoker.

I do not deny that dispositions do indeed require the possession of such generic capacities. My claim is that the ascription of these generic capacities presupposes that persons, objects or substances have some more specific abilities. The ascription of a disposition does presuppose that the object or person to which or to whom we ascribe it has some corresponding ability. Otherwise, our practice to ascribe these capacities would be perfectly empty. Every macroscopic solid object, no matter how hard it is, can break and can be dissolved *simpliciter*. And every such object can turn red, *simpliciter*. The use of such 'can-claims' might be useful in some contexts, but their truth is always grounded in the object's having some more specific abilities.

An agent can have the generic capacity to φ without ever being able to φ. A 'brain in vat' can have the capacity to speak correct English in the sense that it *knows how* to generate grammatically correct English sentences, knows the vocabulary, knows how to initiate a neural process that would 'normally' result in producing English phonemes and so on. But it is never *able* to speak English.

Nonetheless, we can ascribe this capacity to it – to the extent we can – only because it is true that it would talk correct English, if it were placed in a properly functioning human body and tried to speak.

All this shows that modal properties which are abilities are ontologically more fundamental than modal properties which are dispositions. In order to be disposed to φ, an object must possess some ability related to φ-ing. The ascription of dispositions entails the ascription of some corresponding generic capacities or know-hows, which, in turn, entail that objects or persons can acquire some more specific abilities. But dispositions themselves are not to be understood as a special sort of abilities because they do not entail the possession of any specific abilities. We cannot understand what dispositions are with reference to ability ascriptions just as we cannot understand abilities in terms of disposition ascriptions.

Dispositions and laws

Ryle compares sentences that contain disposition terms to the use of statements which express laws. He, like most other philosophers of the early twentieth century, understands laws as statements that have the form of a generalized conditional. In the simplest case, a law states that everything that is F is also G (in other words, *if* something is F, *then* it is also G). Sentences which ascribe dispositions grammatically look like categorical ascriptions (they ascribe a property D to an object *o*), but, in Ryle's view, they implicitly express a conditional about how a person or an object would behave if they were in certain circumstances. The difference between conditionals which express laws and conditionals which are entailed by the ascription of a disposition is that the former must have a universal scope (*everything* that is F is G; or every F that is in C, φs), while the latter concern particular objects (if *x* is in C, then it φs).

I shall argue in this section that Ryle does capture something important when he compares statements ascribing dispositions to statements expressing laws; even if it is a mistake to interpret sentences expressing dispositions as quasi-nomological generalizations about individuals (or kinds of individuals). But an important subclass of dispositions, which I call nomic dispositions, does play an essential role in the explanation of laws. Their role is to explain the distinction between those generalizations which express laws from those which do not.[28]

---

[28] My approach is *not* Humean in the traditional sense – in which Ryle's account certainly seems to be – because I do not reject, as Humeans do, first-order modal properties in ontology. In fact, the whole book is about the significance of such properties in metaphysics. Nonetheless, I agree with

Not every contingently true generalization – or, more precisely, universally quantified conditional – expresses a law. Laws of nature confer a special modal status to their instances in the sense that if o's φ-ing at t is an instance of a law, then it is not merely an accident that o is φ-ing at t. Laws can explain their instances precisely because, whenever an event or a state is their instance, then there is *a sense* in which it is not merely an accident that those events or states occur.

Nonetheless, it is not obvious what that sense is. That all people who participate on a graduation ceremony wear gowns does not seem to be merely accidental. There is a generalization which explains why this happens. If it is asked why John is wearing this fancy black garment this hot July afternoon, it is a very good explanation that he attends a graduation ceremony and that people who attend such ceremony must wear gowns. This generalization also supports the truth of some counterfactual; for if John did not participate in the ceremony, he would certainly not wear a gown. Nonetheless, it is obvious that the generalization about people's wearing gowns on certain occasions does not express a law of nature.

As a first approximation, we might say that a generalization can be nomic only if it is true in virtue of objects' membership in some *natural* kinds. What most laws of *nature* express is that objects *belonging to a natural kind* are disposed either to possess certain features or to display some sort of behaviour in some circumstances. If an object or substance of the kind K is disposed to have a feature F or to display behaviour φ (in circumstances C), then it is not an accident that it has feature F or that it is φ-ing in C because, *given* that it instantiates K, it has the disposition to be F or to φ (in C). Dispositions explain how laws confer a special modal status on their instances.

To illustrate, consider the following two cases. Case [a]: a lizard lives its whole life in a room where the temperature is always 32 degree Celsius. Consequently, it lives its life at a constant body temperature. Case [b]: I, or any other member of the human species, sometimes have fever, and hence our body temperature is sometimes higher than 36.6 degree Celsius. In spite of the invariance of the body temperature of the lizard and the occasional variance of my body temperature, there is no nomological connection between being a lizard and having a *constant*

traditional Humean accounts of laws to the extent that I propose to understand law-statements as generalizations and laws as specific sort of regularities. I shall not consider the major alternative to the Humean account of laws, according to which laws are second-order facts about first-order universals. For the classical exposition of that alternative view see Armstrong (1983). For more on my view on physical laws see Huoranszki (2019).

body temperature; but there is one between me and my having my normal body temperature.

Why is this so? Because whenever I have my normal temperature, this is the manifestation of my genetic disposition to have that temperature in various different thermal environments, whereas for the lizard in my example living its life at a constant temperature is only an accident. What its constant body temperature manifests is its disposition to adjust its body temperature to the temperature of its environment. This suggests that it is an organism's disposition that grounds the nomic connection between membership in a natural kind and the occurrence of certain features or certain kind of behaviour.

But the example also shows that laws do not make the occurrence of their instances necessary. Most laws of nature do not express metaphysically necessary truth. I shall come back to the few possible exceptions later. More importantly, however, even if all laws were metaphysically necessary, this does not mean that the features or events that are their instances occur necessarily. For most nomic generalizations – even some of the 'fundamental' ones – are true only *ceteris paribus*.

At this point, it is important to avoid a frequent misunderstanding. A law is not *ceteris paribus* because it 'has exceptions'. Laws themselves do not have exceptions. What can have exceptions is the generalization that things which instantiate kind K are also Gs. The *law* that '*Ceteris paribus*, Ks are Gs' (or that, *ceteris paribus*, Ks φ in C) has no exception. Neither do such laws state some falsity.[29] *Ceteris paribus* laws express nomic generalizations which allow for some 'exceptions', but which we could not therefore consider as simply false. In fact, some generalizations are such that certain kinds of counterexamples do not undermine our belief in their truth *precisely because* we think that they express a law.

Taking aspirin (meaning the chemical kind *acetylsalicylic acid*) tend to relieve people from headache. If this generalization were false, there would be no point in manufacturing or taking aspirin. But it is not true that aspirin alleviates someone's pain in every case; or that everyone's pain is alleviated by taking it. There are countless factors the occurrence of which can 'falsify' the truth of the strict generalization that whenever someone takes aspirin, she will be relieved from headache. Moreover, trying to enumerate every such factor which can count as non-falsifying counter-instances seems hopeless.

---

[29] Nancy Cartwright once said that laws of nature 'lie'; that is, statements expressing laws are not true. But later she changed her mind and claimed that laws are true only in certain specific situations designed to test them, which she calls 'nomological machines'; see Cartwright (1999), Chapter 3.

Nonetheless, it seems equally implausible to believe that whenever someone takes aspirin, then it is only an accident that she is relieved from headache. The generalization about administering aspirin and pain-alleviation seems to be nomic because it explains why it is not a mere accident that someone's headache is alleviated after taking aspirin *whenever this does happen*.[30]

Thus, laws are generalizations which, on the one hand, are true only *ceteris paribus*, but which, on the other, explain why the occurrence of their instances is not mere accidents. Both of these features of nomic generalizations can be accounted for by objects' nomic dispositions. Such dispositions provide a link between the ascription of abilities and the truth of *ceteris paribus* laws.[31] For instance, aspirin (as a kind of chemical substance) *is disposed* to relieve people from certain kind of pain. It is this dispositional property of (everything which is) aspirin that grounds the law '*Ceteris paribus*, taking aspirin relieves people from (certain kind of) pain'.

The ontological role of dispositional properties is then to ground nomological tendencies. This is compatible with the possibility that, for instance, administering aspirin will in some cases fail to relieve someone from pain. For having the tendency to φ is compatible with the *ability* or liability to not-φ.

Thus, dispositions are not mere abilities. In certain kind of circumstances even a glass of water can, or is able to, relieve someone from her headache. But water is not disposed to relieve people from their headache. For, when drinking water does *not* relieve someone from pain, that is *not* an accident. If taking aspirin fails to do so, then it *is*, in a certain specific sense, an accident. Later, in Chapter 4, I shall try to explain in some more detail what might count as an accident and why. At the moment, we need to note only that whenever it is not an accident that φ happens because its occurrence is an instance of a law, then φ happens or F is displayed because the event (or feature) manifests an object's disposition to φ (or its disposition to display feature F) in the relevant circumstances.

This account of laws and dispositions suggests a close connection between natural kind membership and the possession of certain dispositions. Some laws state that certain kinds of objects in some specific circumstances shall display a certain sort of *behaviour*. Such laws express a connection between some objects

---

[30] This does not mean that philosophers have not argued for both views. Some believe that the *ceteris paribus* conditions can *in principle* be specified, while others deny that *ceteris paribus* generalizations express laws. For a useful summary of the possible interpretations of *ceteris paribus* laws see Reutlinger et al. (2019).
[31] For dispositionalist accounts of laws, which are in certain respects akin to the present view, see Hüttermann (1998) and Bird (2005).

or substance instantiating a kind on the one hand and the tendency to *act* or *interact* with other kind of objects. The ground of these laws is that instantiating some natural kinds entails having certain dispositions.

Other laws state that certain kinds of objects display certain kind of *features*. It is a law that tigers are four-legged animals. What grounds this law is that tigers are *genetically disposed* to have four legs. Thus, it is compatible with the truth of the law that some tigers fail to have four legs. Given the structure of their genes, tigers have the tendency to have four legs. For this reason, it is a law that *ceteris paribus*, if an animal is a tiger, it has four legs.

There are some examples, however, which speak against the close connection between nomic dispositions and natural kind membership. Hardness and being magnetic are nomic dispositions of physical objects. But a piece of glass which is fragile can be hardened, without ceasing to belong to the chemical kind SiO_2. A piece of iron can *become* magnetic without *being* magnetic in virtue of being iron. There are laws concerning the behaviour of hardened glasses and magnetized irons, and such laws can be grounded in some objects' or substances' dispositions, but not *only* because they instantiate a natural kind.

Moreover some, typically *physical*, properties are independent of membership in any specific natural kind. Macroscopic objects can have some electric charge, and each of them has some specific mass, which can ground their dispositions to act, or elicit actions of other objects in a specific manner in certain circumstances, irrespective of which natural kind they exemplify.[32] In fact, it is in this sense that some physical properties might be 'fundamental'. They can be instantiated by any *kind* of material objects or substance. Thus, physical properties seem to be nomic in a special sense.

Consequently, it is worth classifying nomological properties in two distinct groups. A property is nomic if it is associated with a disposition that grounds the laws of nature. Some such properties consider membership in natural kinds; while others are physical properties that can be instantiated across, or independent of, natural kinds. If physics is indeed not a 'special science' – or, in another sense, if physics *is indeed* special – it is so because physical laws are not

---

[32] This observation has been used as an argument against 'dispositional essentialism' by Mumford (2005) and Bird (2007). They also seem to understand mass and charge as properties which 'have powers'. However, as Menzies (2009) notes and I shall argue further in Chapter 7, we do not ascribe powers (or modal properties in general) to properties, we ascribe them to objects. I would interpret such properties as highly abstract properties of matter which depend on objects' more specific abilities. Moreover, the ascription of some such generic dispositions is, in many cases, presupposes natural kind membership. *Chickens* are genetically disposed to have only two legs, *tigers* are not, even if some of them have lost, or have been born with only, two legs.

grounded in dispositions that objects have in virtue of belonging to a natural kind.

It can be objected to this that electrons *must* have negative charge; and hence electrons might be considered as 'fundamental natural kinds'. However, the sense of 'must' here is unclear. Certainly, electrons are now taken to be negatively charged particles *by definition*. If something is positively charged, but otherwise it has the same properties as electrons do (same mass, sort of spins), then that thing is a positron, not an electron with negative charge. But it is also, in a sense, an 'antielectron'. Thus, whether or not an electron *must* have a positive charge is not a matter of natural kind membership, but rather a matter of classification of fundamental particles with reference to their nomic properties (mass, charge, spins, etc.).[33]

Nonetheless the laws that explain how negatively charged particles behave in certain circumstances are grounded in dispositions that objects have in virtue of exemplifying a physical-nomological property. Positively charged particles are disposed to attract negatively charged particles and to repel positively charged ones. For they *tend* to do so, but do not do so in all circumstances. Protons are positively charged particles, but in the presence of (strong residual) nuclear force they are not disposed to repel each other (for the nuclear force completely overwhelms electromagnetic repulsion in the nucleus).[34]

My gravitational mass entails many different dispositions about how my body tends to move, depending on whether I am on the Earth, on an accelerating train, on the Moon or travel in a spaceship. Thus, the instantiation of a physical-nomological property can ground several different, more specific physical laws about the movement of objects considered as physical entities. These more specific laws are all grounded in objects' dispositions to move in a certain manner (with a certain velocity, or with a certain rate of acceleration, in certain direction, etc.) when they are in certain physical circumstances. Such dispositions can also be interpreted as behavioural tendencies of *systems* constituted by me and the earth, or the train, or the Moon, or the spaceship, and so on. I will come back to the issue of systems' dispositions, which seems especially important in physics, in the next chapter.

---

[33] In fact, Paul Dirac, who first predicted positrons on the basis of some theoretical considerations, originally suggested that *electrons* can have both negative and positive charge.

[34] Cartwright provides an example of a physical system in which electrons even tend to attract (!) each other; see Cartwright (1999, 59–61).

Teleological dispositions

Some nomic dispositions are behavioural tendencies that we call *habits*. Cows are ruminant, tigers are predators, stalks are migrant (birds). Habitual behaviour is perhaps the most straightforward example for dispositions grounding behavioural *tendencies*, since habits are actual tendencies of behaviour, where such tendencies must be actually manifested at the level of a single individual. In contrast, every particular object can break only once, even if the *kind* of things to which the individual belong must actually display the relevant behaviour with some frequency.

However, it needs to be emphasized that not every habitual disposition – like being a regular swimmer, smoker or one's tendency to read philosophy regularly – concerns nomic properties. Individuals do not have such dispositions in virtue of belonging to some natural kind (like being *Homo sapiens*). But they share an important feature with habitual dispositions which are nomic. The classification of the sort of behaviour that agents perform when they act habitually depends on whether or not such behaviour manifests their habit. A bird might fly south without migrating. Someone might swim because she has accidentally fallen in water. On that occasion, her swimming does not manifest her disposition to swim regularly.

I am not going to offer any detailed account of the nature of human habits or the logic of the use of terms which are called habituals by linguists here. The former is a topic for philosophy of action; the latter is a problem in linguistics. The reason why I nonetheless need to mention them in a work on modality is to forebear some confusion induced by the different (even if somewhat related) senses in which philosophers use the term 'disposition'. In my view, we have to sharply distinguish between properties that are dispositions in the sense of explaining 'mere habits' as it were and dispositions that express a sort of habitual behaviour which is also nomological. Without such a distinction, we cannot properly understand the modality-grounding role which my theory aims to ascribe to dispositions as nomological properties.

First, whether a piece of behaviour is the manifestation of a nomological disposition depends on whether or not it tends to contribute to attaining some end. Whenever habitual behaviour is nomological, it must be *teleological* too. The relevant ends must be natural, typically biological, like in the case of birds that aim to spend the winter season on warmer lands. Of course, some non-nomological habits like swimming, smoking or satisfying one's interest in philosophy can be teleological as well, since their exercise presupposes conscious, purposive behaviour. But they need not all be. Some 'quasi-intentional' forms of habitual behaviour, like the fiddling with one's pencil while listening to talks or saying 'like' in the middle of the sentences, do not seem to serve any further purpose.[35]

However, every kind of habitual behaviour that is explained with reference to nomic dispositions of some organisms, like the disposition to migrate, ruminate or hunt, are teleological. Most human habits – and at least some animal habits as well – are not nomic because individuals do not have them when they do partly in virtue of belonging to a natural kind. They are dispositions that are typically acquired by 'learning', when such learning involves some form of conditioning (including practice, understood as self-conditioning) and hence actual repetition, and often some prior intention to learn.

It is important to emphasize that not every disposition is teleological just because dispositions are modal properties and hence, as I mentioned in the first chapter, there is a sense in which they are 'intentional' since they are all 'directed at' some event which is their 'manifestation'. The fact that I can (have the ability to) see only if I see *something* does not make that ability teleological. That fragility is 'directed at' the possibility of breaking does not make fragility itself a teleological disposition. I do not explain a body's movement teleologically just by characterizing it in vector rather than in scalar quantities.

In my view – and in this I follow Hegel's very illuminating, partly Aristotle-inspired, discussion of teleology – teleology requires *means-ends connection*.[36] Means can, though need not, be efficient causes, but they must occur, in Aristotelian terminology, for the sake of something. Intentional states are identified with reference to their (intentional) objects, but there is no sense in which they occur 'for the sake of their object'. The same holds for modal properties. We might say that they are 'directed at' some events, but this does not, in itself, make them teleological.

Finally, there is another type of dispositions that I need to briefly mention before closing this chapter. Character traits are one of philosophers' favourite examples of dispositions. From one perspective, this seems very natural. After all, as I mentioned, the original meaning of the word 'disposition' in English meant to express personality traits of agents. From another, however, it seems to be rather misleading to compare character traits either to such nomic

---

[35] I thank one of my referees for this example.
[36] See especially Hegel (1812/2010, 12.163–5).

dispositions like solubility and fragility or to habitual dispositions like being a regular swimmer or smoker.

Given their importance in moral philosophy, there is a huge literature about character traits in philosophy (not to mention in psychology) ever since Plato.[37] Luckily, just as in the case of acquired habits, I only need to mention them here in order to distinguish the problems surrounding them from the issues relevant to the nature of modality grounding, nomological dispositions.

On the one hand, character traits are obviously not nomic dispositions as solubility and fragility are since they vary widely within members of a natural kind and most of them are *somehow* acquired. Whether they are acquired by conditioning or by some other kind of learning procedure has puzzled philosophers ever since Aristotle. But it seems to be certain that character traits are not mere habits in the sense in which being a smoker or being a regular swimmer are. It is true that character traits are often manifested by agents' intentional actions, and hence their ascription can explain some goal directed, teleological behaviour. Nonetheless, a behaviour does not manifest an agent's character *only because* its performance tends to contribute to certain goals. Irascible people tend to react violently to what they perceive as an offence, but not because they have a specific goal which such type of behaviour tends to satisfy.

Moreover, what many philosophers today, especially after Ryle, mean by their claim that character traits are dispositions is that if someone has a certain character trait, then she is disposed to *act* in certain ways. But this seems to be wrong. Agents' actions themselves need not manifest their character since character traits are not manifested merely by tendencies of overt behaviour. Their manifestation also involves persons' emotional reactions and some affective states that occur concomitant with their actions.[38]

How people can tolerate frustration, respond to the success of others, endure pain or challenges depends on their traits, but such properties are not always straightforwardly manifested in overt behaviour. Behavioural tendencies are often not sufficient for identifying emotional ones, because there need not be a one-to-one correspondence between agents' emotions and their actions.

In general, the logic of trait-attribution and how such attributions can explain behaviour is a difficult theoretical issue in philosophical psychology and

---

[37] I say more of the problem of traits and their relation to agents' abilities and dispositions in Huoranszki (2011, 170–5).
[38] A dentist doing her work often causes inevitably some pain. The dentist, who can cope with this, and the dentist, who positively enjoys this, are obviously different in character even if they perform the same type of actions in the relevantly same type of circumstances.

psychology itself.[39] Given the complexities of trait attribution and the difficulty to find out about people's real character, it is even somewhat surprising why, etymological reasons aside, philosophers mention character traits routinely as paradigmatic examples of dispositions.

Hence all I want to say about character attributions here is that even if traits are obviously linked somehow to what people are disposed to do, or how they are disposed to react emotionally or cognitively to certain situations, the sense in which such traits are 'dispositional' is fundamentally different from the sense in which properties like fragility or being a predator are dispositional. Their nature and manifestation is a complex problem, which requires an entirely separate discussion related to the study of intentional human behaviour. For this reason, I shall not discuss the specific issue of traits as dispositions here.

---

[39] For an interesting attempt to analyse character traits as disposition see Butler (1988). For further considerations about the question whether we should interpret traits as dispositions see Alvarez (2017).

3

Specificity and extrinsicness

In the previous chapter I argued that abilities and dispositions are different modal properties because they play different roles in ontology. In this chapter, I shall discuss the issue of specificity of abilities and dispositions. Specificity is a matter of degree. Modal properties can be more or less specific. But they are not specific only in the sense in which some physical properties (pressure, volume, masses, forces, etc.) are. Unlike such physical properties, abilities and dispositions are not only *quantitatively* but also *qualitatively* specifiable.

In the next chapter, I shall argue that it is for this reason that we cannot dismiss counterfactual conditionals in the specification of modal properties. The present chapter's topic is specificity rather than specification; that is to say, ontology rather than semantics. But it provides the background for the next chapter's account about the connection between the ascription of modal properties, like abilities and dispositions, and counterfactual conditionals.

As we shall see, qualitative specificity and generality are closely related to the distinction between intrinsic and extrinsic properties. I shall argue that the intrinsicness and extrinsicness of modal properties are to a large extent a matter of their generality and specificity. One interesting consequence of this connection is that, since qualitative specificity is gradable, so must be the intrinsicness and extrinsicness of modal properties.[1]

Specificity plays a fundamental role in my proposed account of contingency. Possibilities can be more or less abstract. In the possible world approach, this is expressed by 'a world's distance' from actuality. The 'farer' a world is, the more abstract the possibilities it represents are. I shall suggest an alternative way to understand specificity and abstractness. Possibilities can be more or less abstract because the modal properties that ground them can be more or less generic. The more specific an ability is, the less abstract the possibility which it grounds will be.

---

[1] For a similar argument about gradable extrinsicness see Barbara Vetter (2013).

Further, I shall argue that abilities, unlike dispositions, can be *maximally* specific. Such abilities ground the least abstract contingent possibilities. In contrast, dispositions must be more generic – though they are also generic to a variable degree – otherwise they could not fulfil their role to ground behavioural tendencies. As we shall see, this is the reason why many traditional accounts of dispositions take them to be necessarily intrinsic. Qualitative specificity is often extrinsic, but dispositions are rarely qualitatively specific. For this reason, it is natural to consider them to be intrinsic.

Specificity

Suppose we want to specify the property 'toxic' with the help of some explicitly dispositional locution. As a first step, we can use the simple paraphrase: a substance is toxic if and only if it is disposed to cause the death of those who ingest a certain amount of it.

But the simple specification needs to be hedged. Arsenic is toxic, even if it is not disposed to poison *everyone* who ingests it. Arsenic has an antidote. Thus, it seems that the explicit specification of 'toxic' must include some reference to the absence of such antidotes. However, introducing this extra condition leads to an ambiguity when we try to apply it in specific cases.[2]

Suppose that we modify our dispositional locution like this: a substance is toxic if it is disposed to poison those who ingest it without taking antidotes. Suppose that Jill ingests some sample of arsenic (for simplicity's sake, name it S_A) and she also takes antidotes. Is then S_A (the specific sample of arsenic) toxic or not?

According to one possible reading, the specific sample S_A is *not* toxic, since it is not disposed to poison Jill who ingests it; and nothing can be toxic if it is not disposed to poison someone who ingests it. But since Jill who ingests S_A takes antidotes, S_A is not disposed to poison her. Hence, the specific sample of arsenic S_A in the circumstances in which Jill ingests it is *not* toxic.

According to another, seemingly more plausible, reading, S_A is toxic. After all, arsenic is a toxin. The relevant sample of arsenic is toxic because it is disposed to poison those who ingest it and *do not* take antidotes. The reaction of those who do take antidotes is irrelevant to the specification of arsenic's biochemical

---

[2] This problem was first explored by Bird (1998, 231).

dispositions. This means that a sample of substance can be toxic even if it is *not* disposed to poison someone who ingests it.

To illustrate the dilemma from the other direction, as it were, consider the following case. Sodium is a corrosive substance. It is disposed to burn body tissue with which it comes into contact. But when sodium is ingested in the form of sodium chloride (common salt), it is not disposed to cause any damage to the body tissue. In fact, consuming it in the appropriate quantity is necessary to sustain health.

Now consider Jack who adds salt to his salad and then eats it. He contacts sodium then. Is the sample of sodium he is contacting disposed to damage his body? It is not disposed to harm when it is ingested and bonded with chloride. Yet, sodium is certainly disposed to burn his body tissue if it is contacted in circumstances in which it is *not* bonded with chloride.

Most things are disposed to display a certain kind of behaviour or cause certain changes only in some specific circumstances. Does this make the very disposition that we aim to specify extrinsic to the object to which we ascribe it? It seems that the answer depends on how we interpret explicitly dispositional locutions. According to one interpretation, '*o* has the disposition D in C if and only if it is *disposed to φ in C*'; according to another, '*o* has the disposition D if and only if, *in some specific circumstances C*, it is *disposed to φ*'. Unfortunately, it seems that both interpretations lead to some counterintuitive consequences.[3]

If the first interpretation is the correct one, then dispositions are indeed extrinsic. The object has the disposition *only* in circumstances C. This would mean that a sample of substance is toxic *only* in the circumstances in which antidotes are absent. And a substance is corrosive *only* when it is digested (or touched) in unbonded form. This may sound counterintuitive. A sample of arsenic seems to be toxic, and sodium seems to be a corrosive substance, even in the circumstances in which they fail to harm those who touch or ingest them.

The second interpretation does not entail that dispositions are extrinsic. Having a disposition D entails only that there are circumstances in which objects or substances are disposed to display a certain sort of characteristic behaviour. Arsenic is toxic because in circumstances in which antidotes are not taken, it is disposed to poison people; and sodium is corrosive because in circumstances in which it is not bonded, it is disposed to damage living tissue. However, if this is

---

[3] See Bird (1998). I slightly modify Bird's own presentation of the problem, since his focus is David Lewis's conditional analysis of dispositions, some interesting features of which I shall discuss later.

the correct interpretation of being toxic or being corrosive, then too many things turn out to be toxic and corrosive.

In some appropriately specified circumstances, many things can poison people: salt, peanuts, vitamin D or some extreme amount of water. If the second interpretation were correct, we should regard salt, peanuts, vitamin D or water toxic because, in some circumstances, they are disposed to poison those who ingest them. Similarly, many kinds of elements should be regarded as corrodent, since, in the circumstances in which they are bonded in acids, they are disposed to damage body tissue. But this is implausible. Dispositions ground statistical tendencies, and it is false that ingesting peanuts tends to poison people or inhaling nitrogen tends to corrode their lungs (just because nitric acid would do so).

What this problem shows, first about dispositions, is that dispositions can be identified only with reference to how substances and objects *tend* to behave (or what they *tend* to cause) in some circumstances; that something *can* φ in *some* circumstances is not sufficient to ascribe the disposition to φ to it. However, second, the problem illustrates further the difference between dispositions and abilities. For what seems to be an implausible conclusion about dispositions is not implausible about objects' (or substances') abilities.

Even if it does indeed sound false that salt, peanut, vitamin D or water are toxic in the sense that they are *disposed* to poison those who ingest them, it is very plausible – in fact, it is true – that they *are able* to poison (or at least harm) those who ingest them. Many abilities of chemical substances have a similar characteristic. A substance might be able to φ when it is unbonded, but not able to φ when it is bonded; or the other way around, it might be able to ψ when it is bonded, but not able to ψ when unbonded. More generally – and independent of the illustrative chemical examples – we can identify *different*, though not necessarily distinct, abilities of substances or objects depending on how we interpret the role that certain extrinsic conditions play in their specification.[4]

Abilities, like many other properties, might be more or less generic. A more generic modal property can ground more abstract possibilities. A concrete sample of sodium S_N which is bonded in salt and mixed in an otherwise healthy food does not have the ability at t to harm its consumer's body tissue. Yet, S_N has the more generic ability even at t to harm someone who eats (touches) it in unbonded form. This latter possibility is more abstract than the former one,

---

[4] This has been already noted, in an entirely different context, by Ernest Sosa in Sosa (1993, 317).

since it is grounded in S_N's more generic ability, which we can ascribe if we *abstract away* some features of the actual circumstances.

Philosophers use the notion of 'abstract' at least in two different senses. In one sense, something is abstract if it is not part of the spatiotemporal world and, correspondingly, does not exercise, or is not exposed to, any causal influence. Numbers, meanings, constitutions or fictions are meant to be abstract in this sense.

In another sense, something is abstract if it is *abstracted from* the concrete. Nemo is a dachshund. But Nemo is also a dog. Being a dog is his more generic, and hence more abstract, property than being a dachshund because *fewer specific conditions need to be satisfied* in order for an animal to be a dog than in order for it to be a dachshund. For this reason, something *can* be a dog without being a dachshund. And further, something can be an animal without being a dog. Thus, being an animal is even more generic and hence more abstract property than being a dog.

Being abstract in the latter sense, like being generic, is a matter of degree. The level of generality of a modal property is determined by how many conditions must be *actually* satisfied in order for something to possess it at a time. In a sense, no property is 'concrete'; only their instances are.[5] But properties can be more or less specific.

Correspondingly, contingent possibilities – unlike, for instance, mathematical possibilities – can be more or less abstract. The more specific a property is, the less abstract possibility it grounds. In this sense, it is more adequate to contrast the abstract with the *specific* than to contrast it with the concrete. Alternatively – as Hegel seems to hold – getting more specific just means in this context getting more concrete.[6] In either way, contingent possibilities at various levels of abstraction are grounded in the more or less generic modal properties of objects, persons, chemical substances or ensembles of particulars.

Peanuts are able to poison *some people* who ingest them. This is a rather generic ability, which grounds an abstract possibility. Nonetheless, the very same peanuts still lack the specific ability to poison me, at least in normal circumstances and in

---

[5] Unless specific property-instances like tropes are taken to be properties as well. Tropes are indeed concrete and determinate, but for this reason, I would not classify them as properties.

[6] The first modern philosopher who understands abstractness as a matter of degree was of course Locke. But Locke was interested only in 'ideas'; that is to say, representations. It was Hegel who first interpreted the gradeability of abstractness/concreteness ontologically and who gave a prominent role to this in his philosophy. Abstraction/concretization plays an important role also in philosophy of science, first in Nowak (1979) (though he calls the process of abstraction 'idealization'), and then in Cartwright (1989), Chapter 5.

my present physical condition. This is compatible with these peanuts' possessing the more abstract ability to poison those who are allergic to peanuts.

There are various ways in which we can initially identify specific abilities. The simplest one is to use demonstratives. For instance, we can say that *this bit* of arsenic material does not have *the specific property* to poison *these* people *now*. Either because they have taken antidotes or because they are going to die before the arsenic has time to develop its effect; or because it is so safely packed that the people who have access to it are not even able to ingest it; or perhaps for some other more or less fancy reason. But even when *that* bit of arsenic in *those* circumstances is not able to poison *these* people, it has the *more abstract property* of being able to poison some. And arsenic also has the ability to poison *these* people in the nonactual circumstances in which they do not take antidotes, and so on.

This suggests that it is *necessary to explicitly specify* the conditions in which arsenic would exercise its ability only when we want to ascribe some more generic ability that grounds some more abstract possibility. Thus, the problem of explicit specification of *all* relevant factors that are necessary for the ability to be exercised arises only when we want to identify some very generic ability. Otherwise, the statement that (in our example) *this* bit of arsenic is able to poison *these* people if they ingest it in the *actual* circumstances is simply false. This has some important consequence to the interpretation of the so-called conditional analyses of modal properties, which I shall discuss in detail in the next chapter.

Since abilities can be more or less specific, they are also related to each other as determinate and determinable properties. Having a specific ability entails having the more generic one. If *this* bit of arsenic is able to poison *this* person, then it must be able to poison some (sort of) people in some (sort of) circumstances.

It is worth mentioning though that in this respect there is an important difference between abilities and other properties.[7] In typical cases, although the instantiation of a determinable (like being coloured) does not entail the instantiation of one specific determinate of that determinable (like being red), it does require that *some* determinate property of that determinable be instantiated.[8] In this sense, every determinable is a modal property. Nothing can be coloured,

---

[7] I shall discuss the distinction between modal and nonmodal properties in the last chapter. Those who are sceptical about the existence of nonmodal properties can read the next paragraphs as introducing a distinction within the class of modal properties.
[8] Although even this claim might be contentious. Funkhouser (2006) argues that the instantiation of determinables entails some determinate; for a criticism see Sanford (2016). This debate is related to the general issue of metaphysical indeterminacy, which is not my present concern. See also Wilson (2012, 2021).

without having some determinate colour; and nothing can be angular without being *n*-lateral (when *n* is a specific positive integer). The relation between specific and the corresponding more generic abilities is different in this respect.

An object cannot have a more determinate ability without having some corresponding determinable. After all, in many cases the determinable ability is *abstracted from* the more determinate ones. But an object can have the more generic ability without having a specific, more determinate one. Objects can have more specific abilities only if certain conditions are *actually* satisfied, but they can have the more generic ones even when those conditions do not actually obtain.

On the other hand, some abilities are *maximally specific*.[9] Importantly, when we ascribe an ability that is maximally specific, the problem of how to identify all the relevant conditions which need to be satisfied for an object to possess the relevant specific ability does not arise. Suppose I say that the mug in front of me is able to break now in the sense that it would break in circumstances which differ from the actual one only to the extent that, in the alternative circumstances, the mug falls from the table. I do believe that my mug does have that specific ability. Of course, it might not. Perhaps there are some such factors actually present which undermine the mug's ability; and then it has only the more abstract ability to break in the circumstances in which it falls and in which these factors, actually present, are absent. But not knowing for certain whether or not this mug has that most specific ability does not mean that it must fail to have it.

I say that such maximally specific unexercised abilities are (modal) properties of objects. But does it make sense to ascribe such properties? One might worry that maximally specific abilities cannot be 'real properties' because 'real properties' must be 'sparse' and 'natural'. And maximally specific abilities are obviously very 'abundant' properties and hence they are 'unnatural'.[10] One might object that even if we can use fancy *predicates* as (something) 'is such that it is able at t to φ at t'' in C if every condition necessary for φ-ing, but unspecified by C, is actually satisfied' if we want, why should we grant that this is a 'genuine' property of an object and not just a (strange) way of talking?

My answer is that all depends on what we mean by 'sparseness'. Consider laws of physics, which are typically functional laws expressed with the help of equations;

---

[9] If one prefers reasoning 'with worlds', one can say that a maximally specific ability grounds possibilities represented by worlds that are the closest, that is to say, most similar, to our world.
[10] About the concept of 'sparseness' and 'abundance' of properties see Lewis (1983). As I understand him, Lewis does not deny the existence of abundant properties; he only says that they are not 'natural'. In my view, however, naturalness itself has little to do with parsimony. For an argument that modal properties need not be sparse, which is different from mine, see McKitrick (2003).

for instance, F=*ma* (force is equal to the product of mass and acceleration) or P=*f/s* (pressure is equal to the ratio of force and the area on which it is exerted). Such laws concern *infinitely many specific properties of objects*, given that their variables can take infinitely many values. But no one would complain that specific physical properties are not 'natural' because we can ascribe to objects infinitely many determinate masses, forces, areas of surface and so on.

Moreover, the ways we characterize such physical properties are almost always more or less specific. There is hardly any physical parameter the value of which can be *measured* with maximal exactitude. But no one would claim that this undermines the truth of physical laws. And it would be entirely irrational to deny that physical objects have the more or less specific property of mass, volume, surface and so on just because they are almost as 'abundant' as the number of actual physical objects are.

Exactly the same holds for the modal properties of objects. The only significant difference is that the specification of such properties is not always done in quantitative terms. If our project is to ground contingent possibilities in objects' modal properties, we have good reason to assume that properties are abundant, since possibilities are abundant too. We ascribe abilities in order to identify what *things can do* and, pragmatically, what *we can do with them*. And ascribing maximally specific abilities is entirely natural in this context.

When we need to decide what to do, first we want to know our real options; that is to say, what we can do, or *whether or not we have the ability* to do something, *in the specific circumstances* in which we need to make a choice. Suppose I sit in a firmly locked room without keys and so on. Then I'm not able to leave it. I do not have the specific ability to leave *that room at that time*. I do have a generic ability to leave a room with unlocked doors or doors to which I have the key (because I am able to move my limbs, see exits, use keys, etc.). But when the practical question about my next action arises in a particular situation, my concern is whether I can do this or that *at that specific moment and in those circumstances*. I assume that I have (or lack) a specific ability in the given circumstances, even if I cannot explicitly identify all the conditions that need to be actually satisfied in order to have that specific ability.[11]

Similarly, the ascription of specific abilities can also play a role in the explanation of what has actually happened on a particular occasion. Suppose we want to explain why a road accident happened. We might say that it happened

---

[11] I say more about this in the specific context of human abilities, and about how all this is related to the traditional problem of free will and the ascription of responsibility, in Huoranszki (2011).

because a vehicle was unable to halt quickly enough in the circumstances. Perhaps it was unable to halt because it went too fast given the actual conditions of road; perhaps the driver was unable to react fast enough because she was too tired, and so on. Whatever the reason was, the accident is explained by the lack of some ability that an object or a person had or failed to have *in those specific circumstances*.

Importantly, all this does not mean that objects cannot have more generic *abilities* in those actual circumstances in which they lack some related specific one. According to what is called a 'butterfly effect', my far-off sneeze can bring about such physical disturbances that result in the shattering of a glass. Should we say then that I have *the disposition* to break windows by sneezing? This sounds bizarre.[12]

But it is not at all wrong to claim that I have the *ability* to break a glass by a far-off sneeze. That ability is, of course, extremely generic in the sense that it would be exercised only in circumstances in which countless conditions were different from the actual. This generic ability grounds a very abstract 'faraway' possibility. But, if butterfly effects are indeed physically possible, then that very abstract 'faraway' possibility must exist as well; and what grounds it is a property, which is my very generic ability.

Here is another, nonphysical example. Can I be the next prime minister of Hungary? As things stand now, I certainly do not have that specific ability. Nonetheless, I do have the (much) more generic ability that if several nonactual conditions were satisfied, I would become prime minister. Nothing makes it physically, legally or metaphysically impossible for me to become a candidate and to be elected. Of course, as far as my actual intentions and actions are concerned, this is an uninterestingly abstract 'faraway' possibility, almost as remote as the possibility of my breaking a far-off window by a sneeze is. Yet, it is *a* possibility.

Most properties of objects are uninteresting for us in most contexts. Yet, objects still have them. No one is interested in the precise number of hairs on the heads of British MPs who voted against a proposal on a specific occasion, since that number does not explain anything. Nonetheless, there must be such a number. Analogously, in the actual circumstances in which we live, probably none of us has the *specific* ability to break glasses by far-off sneezes; and hence, in most contexts, we are not interested in ascribing that very generic ability to anyone. Nonetheless, if current physics is right, each of us has that ability, which grounds the very abstract ('faraway') possibility to perform such feats.

---

[12] See Bird (1998, 231).

Extrinsic abilities and abstraction

It follows from this account of abilities that most specific abilities are extrinsic. Many philosophers find this objectionable. It is a widely – though not universally – granted assumption that dispositions and powers need to be intrinsic properties of objects.[13] But if abilities ground more or less abstract possibilities in the sense explained earlier, then we have good reasons to believe that many of them are extrinsic. For, in most cases, whether we can correctly ascribe some abilities depends on the *actual* presence or absence of some factors, which are extrinsic to the object(s) or substance that possesses the ability.

The view that modal properties must be intrinsic is especially common among those philosophers who are realist about such properties; that is, among those who regard modal properties as irreducible, first-order properties of objects. This might be so because many philosophers believe that realism about some sort of properties is incompatible with the extrinsicness of those properties.

Here is an argument often used to support this claim. Xantippe can become a widow and Socrates can become shorter than Plato. But becoming a widow and becoming shorter than someone else are only 'pseudo' (or 'Cambridge')-changes *because* 'being a widow' and 'being shorter than Plato' are 'merely' extrinsic properties of Xantippe and Socrates. The 'real', 'genuine' changes are Socrates's death and Plato's growth, since those are changes of properties that are intrinsic to their bearers.[14]

Similarly, one might argue that extrinsic abilities (or dispositions) cannot be 'genuine' properties. If objects' abilities can ground possibilities at all, such abilities must be intrinsic to them. This seems to be a serious challenge for the project to understand contingency, and hence possibilities of the less abstract sort, in terms of ascribing abilities to objects. For, as I have just argued, abilities which ground possibilities relevant in the context of practical and theoretical reasoning are usually specific and, hence, extrinsic.

However, the intuition that extrinsic properties cannot be 'real' or 'genuine' is based only on the use of some arbitrarily chosen examples. The property of

---

[13] Though Popper does not use the notion of 'extrinsicness', arguably, his propensity interpretation of probability entails it, since in his view propensity-dispositions can be ascribed to object(s) only in the whole experimental set-up. See Popper (1957, 67). As mentioned earlier, Jennifer McKitrick argued convincingly that some dispositions – vulnerability, visibility or weight – seem to be extrinsic to the objects to which we ascribe them; see McKitrick op. cit. The extrinsicness of dispositions has already been assumed, though not explicitly argued for, in Smith's account of dispositions Smith (1977). More recently Barbara Vetter argued in detail for extrinsic abilities in Vetter (2013). Much that is said in this chapter is congenial with the views developed in that paper.

[14] For further discussion with specific reference to modal properties see Shoemaker (1980).

'being a widow' depends on what has happened to someone's spouse; and 'being shorter than Michael Jordan' depends on the height of those who are concerned. This shows that the instantiation of some properties can be entailed by the instantiation of some other(s). But this does not mean that the properties on which they depend must be intrinsic.

In fact, as I shall argue, many abilities and dispositions are extrinsic at least in one sense of the intrinsic/extrinsic distinction. Thus, although there is a connection between the intrinsicness and generality of abilities in the sense that the more generic properties are also the more intrinsic ones, being extrinsic does not make a property 'less real' or 'not genuine'.

Water can vaporize or solidify, and it can dissolve salt and sugar. Thus, water has the ability to vaporize and to freeze, as well as the ability to become salted or sweetened. That water has the ability to freeze seems to be an intrinsic property of water. Of course, freezing requires being at a certain temperature, but there is no water (or chunk of matter) without temperature; hence, if there is water, it must also have the generic ability to freeze. But water has the ability to dissolve sugar only in circumstances in which sucrose molecules evolve in the history of the universe.[15]

We might insist that even if there were no concrete instances of sucrose in our world, water might have the more abstract ability to dissolve substances with certain chemical properties characterized very generally, of which our *actual* sugar is an instant. But as long as there is no sucrose in the world, there is no reason to ascribe the ability to dissolve sugar to water in our world. Or at least, so I shall argue, in the next section of this chapter. Before I do so, however, I wish to end this one with a few general remarks about the debates over the intrinsicness of modal properties.

Patently, whether or not we regard some abilities extrinsic depends partly on how we understand the distinction between extrinsic and intrinsic properties in general. It is usually assumed that we must have some more or less clear 'pre-existing intuition' about extrinsicness and intrinsicness, on which a general account of the distinction must rest. However, even if this might be right about some cases like 'being shorter than Michael Jordan' and 'being a widow', it does not seem to be right in the context of modal properties. We do not seem to have any universally granted 'pre-existing' intuition about such properties' intrinsicness or extrinsicness. For instance, some philosophers take dispositions

---

[15] Common sugar's molecular formula is $C_{12}H_{22}O_{11}$; it is easy to imagine histories in which such complex chemical substances do *not* evolve.

to be *obviously* intrinsic, whereas others believe that they are *obviously* extrinsic properties.[16]

Moreover, the claim that modal properties are intrinsic does not, in itself, entail any commitment to their being genuine, unreducible properties. As we shall see in the next chapter, David Lewis, who rejects first-order modal properties and explicitly proposes a 'reductive' analysis of dispositions, still holds that they *must* be intrinsic.[17] In fact, as we shall also see, it is a crucial assumption of many anti-realist, reductivist analyses of dispositions – and modal properties in general – that they are intrinsic to their bearers. Thus, insisting that modal properties *must* be intrinsic is certainly not, in itself, an argument for realism.

I shall argue then that what makes an ability extrinsic is the fact that *its possession is sensitive to* changes in its bearer's environment. If the environment changes in certain ways, so does the object or substance: it can lose and gain some abilities. I shall also argue that such sensitivity is *a matter of degree*. The instantiation of maximally specific abilities is usually very sensitive to their bearer's circumstances. But, again, there are abilities at various levels of generality, which can ground more or less abstract possibilities.

The more generic a modal property is, the less sensitive it is to the alterations of actual circumstances. In one sense – but only in one sense – the level of generality of an ability (and hence the level of abstractness of the corresponding possibility) is determined by how sensitive the possession of an object's, person's or a substance's ability to the changes of its actual environment is.[18]

Since the more generic abilities are the least sensitive to the changes of their bearers' actual environment, there is a sense in which we can indeed classify such abilities as *intrinsic* properties of objects. But while there are *maximally specific* abilities which are most often extrinsic, there are not *maximally abstract* physical abilities which objects can have irrespective of any factors in their actual environment.[19] I shall continue to argue herein that there is an important sense in which even abilities of the most generic sort can be sensitive to some features

---

[16] As David Lewis and Rea Langton already noted, see Lewis and Langton (1998/1999, 123).
[17] See Lewis (1997/1999, 147).
[18] This is obviously not the only determinant of generality since modal properties' generality is also determined by the types of events which manifest them. The ability to swim is more generic than the ability to butterfly; and the ability to butterfly is less specific than the ability to butterfly a hundred metres within two minutes. Thus, abilities can be more or less determinate along two dimensions: first, according to the type of event that are exercised when they are manifested; and second, according to the circumstances that must actually obtain in order for an object to possess the relevant abilities.
[19] My arguments concern only physical abilities. Whether or not any mental ability can be maximally intrinsic is a contentious issue in the philosophy of mind, which is beyond my present concern.

of their bearer's actual environment. For physical abilities are abilities *in the world as a system*.

Intrinsicness and duplications

Intuitively, my height is my intrinsic property, whereas my being shorter than the Eiffel Tower, or being at a certain distance from it, is an extrinsic one. Roughly, whether or not I have an intrinsic property depends only on the way in which I actually am, whereas the possession of extrinsic properties also depends on where and how other things are. These are indeed only rough intuitions about intrinsicness, and it has turned out to be tremendously difficult to spell them out more precisely.[20]

This is no accident. For it is unclear why 'the way I actually am' depends only on properties that are, in some sense, intrinsic to me. Consider a popular proposal. A property of mine is intrinsic if a lonely duplicate of me has it too. My duplicate is lonely, if he exists in an environment where nothing else exists. This explains the sense in which **my being six metres away from a rhododendron** is my extrinsic property. My lonely duplicate does not have it. But is then **not being six metres away from a rhododendron** extrinsic too? It seems that it should be, since it is a property in respect of which an object can change merely in virtue of the change of the location of a rhododendron in its environment. But this is a property which 'my lonely duplicate' can, in fact must, have.[21]

This shows that 'perfect duplication' might not be a very helpful test for understanding which properties are extrinsic. And there is a reason why it is not. Intuitions about duplications tell us more about the *object* duplicated than about the nature of its properties. Intrinsicness is often associated with *essentiality*. But intrinsicness is not the same as essentiality; duplication-intuitions might reveal more about the latter than about the former. My distance from rhododendrons is certainly not essential to me. But for me to exist as a living being it is essential that I have access to oxygen from my environment. That a property is extrinsic does not entail that it must be inessential.

---

[20] About the extrinsic/intrinsic distinction more generally see, among others, Humberstone (1996), Vallentyne (1997), Langton and Lewis (1999/1998), Yablo (1999), and Marshall and Weatherstone 2018). I borrow the example in the following paragraph from Humberstone (1996).
[21] The idea that intrinsicness should be understood in terms of duplication is implicit in Kim (1982/1993) and has been defended against a similar objection in Langton and Lewis (1999/1998). But their account does not answer my main concern about the notion of duplication.

Before I proceed, I would like to address a difficulty which my argument earlier allegedly faces.[22] The argument seems to presuppose the existence of 'negative properties'. But many philosophers, following David Armstrong, reject such properties. This would indeed be a serious objection to the argument, if the rejection of 'negative properties' were sufficiently well motivated and, further, if the distinction between 'negative' and 'positive' properties withstood scrutiny. But it is not well motivated, partly because the distinction is arbitrary. It seems to me that the idea of 'negative properties' is the result of an illegitimate move from correct considerations about negating sentences to incorrect considerations about properties.

Suppose that statement p is contingently true. Then p states a fact. If p is true, then its denial, *not-p* must be false. But what does *not-p* state? A 'non-fact'? This sounds absurd; 'non-facts' are not entities in the world. However, negating a true proposition is one thing; 'negating a property', if this is possible at all, is quite another. There is nothing absurd in claiming that if a rose is not red, then it is non-red. In fact, this seems not only natural but quite right.

Moreover, there are many reasons to take allegedly 'negative properties' seriously in both science and ontology. Science requires negative properties since exclusion laws cannot be formulated without ascribing a negative property to a system, and some exclusion laws (like Pauli's exclusion principle) are quite fundamental. Other important properties of physics, like being an isolated system, cannot be understood without the ascription of negative properties. The same is true about chemistry. Some metals are *in*soluble (i.e. not soluble) in certain acids; others are not insoluble (that is: they are soluble). There is no real motive to deny that insolubility in acids of type A (i.e. having the disposition *not* to dissolve in acids of type A) is a less decent property than solubility is. The same is true about biology; immunity seems to be the 'negative property' of *not* being susceptible to certain kind(s) of disease.

Metaphysics requires such allegedly 'negative' properties, because without them we could not give a proper account of the nature of such entities as shadows, holes, limits (i.e. any 'nonsubstantial' spatiotemporal particulars) and we cannot make sense of the idea of privation in general. Further, the ascription of some determinates entails the ascription of negative determinates of the same determinable, while the ascription of some others does not. If something is green, then it is non-red, non-blue, and so on; but if something is sour it can still

---

[22] Thanks to one of my reviewers for raising this issue.

be sweet. If there are no negative properties, we cannot express this important exclusion feature of determinates of certain determinables.

Most importantly, however, the very distinction between 'negative' and 'positive' properties is arbitrary. Some properties are contraries. If they are, it follows that if something is not F, then it is a non-F. But a thing's non-F-ness is its property just as its F-ness is. Consider opacity. Is it the 'negative' property of *not letting* radiation penetrate a substance? Or is it a 'positive' property of being able to *obstruct* the propagation of radiation? Conversely, consider transparency: is it a 'positive' property (disposition) of allowing light to pass through a material or the 'negative' property of *not* obstructing the propagation of radiation? If opacity is positive, is then transparency negative? Or perhaps the other way around? Who knows? And why should we care? It is so easy to pretend that we understand what we mean when we say that 'F' is a property, but 'non-F' is not a property, only so long as we do not bother to fill in, as it were, this abstract 'principle' with real content.

Finally, let me note that in the standard characterization of intrinsicness which I discuss here, philosophers rely on the idea of *lonely* duplicates. Loneliness, if anything, seems to be a 'negative property': it means *not* being surrounded by anything. So, if there are no 'negative properties', the characterization of intrinsicness under consideration is already doomed – no more work needs to be done. However, even if I am critical of this specific account of intrinsicness, I do not want to reject it on the ground that it assumes 'negative properties'. Loneliness is 'negative' in some sense, positive in some other.

Be this as it may, I want to focus here on the issue about the extrinsicness of abilities since I am interested in here in the nature of modal properties, not the concept of extrinsicness in general.[23] And that problem is largely independent of the issue about 'negative properties'.

The way in which I actually am now includes my ability to drive cars. Is this ability an intrinsic ability of mine? On the one hand, it may seem obvious that it is. Think of my perfect duplicate in an environment in which there are no cars, no roads, just me. Intuitively, my perfect duplicate must 'carry over' the ability to drive cars, which I actually have, into his impoverished environment.

On the other hand, however, why should we assume that the intuition that my duplicate 'carries over' this property is a reason to take the property to be intrinsic in the first place? By the same reasoning, we can say that the property

---

[23] In fact, I tend to agree with Humberstone that there is no universal distinction between intrinsic and extrinsic properties, but the criteria we use to distinguish them depend on which work the distinction aims to do in some specific context. See Humberstone (1996).

of my being shorter than the Eiffel Tower can be 'carried over' to an imagined environment in which there is no Eiffel Tower; or in which the Eiffel-(tower?) is only one metre tall. After all, my property of **being shorter than the Eiffel Tower actually is** will *not* change in the imagined circumstances. So, what reason do we have to deny that my lonely duplicates can have *that* property?

This shows that intuitions about which properties perfect duplicates must – or must not – share are intuitions about the 'perfectness' of duplication, not intuitions about the nature of properties. From the duplication cases we might learn something about the nature of a property only derivatively; but what we learn does not help explain the intuitive distinction between extrinsic and intrinsic modal properties.

If we have the intuition that my duplicate has the ability to drive cars even in car-less circumstances, we might conclude that there is a sense in which that ability is intrinsic. But we can use the duplication scenarios, with equal plausibility as I shall argue herein, to support the claim that *perfect duplicates can share properties that are extrinsic*. Both specific and generic abilities can be extrinsic. But while my duplicate without any access to cars would lose the specific ability to drive one, it would not lose the generic ability to drive cars. The question whether my lonely duplicate can have the ability to drive cars is not a question about the intrinsicness of that property; rather, it is a question about what it would mean to duplicate *me*.

It makes sense to ascribe the ability to drive cars to my duplicate in the complete absence of cars and roads not because that ability is intrinsic, but because duplication is not a symmetric relation. A duplication is a *copy* of something that is the *original*, but the original is not a duplication of its own copy. In that sense, a duplicate can possess properties *in virtue of* the object of which it is a duplicate; but a duplicated object cannot have a property in virtue of the duplicate's having it.[24]

A 'lonely duplicate' of one of Degas's paintings can have the property of being the duplicate of a valuable object. But a lonely painting without anyone to value it cannot be a valuable object. Analogously, my lonely duplicate can have the generic ability to drive cars in virtue of *me* actually having it; even if he obviously lacks the more specific one, which requires access to some cars and roads.

All this, however, does not help understand the *nature* of the property that both I and my duplicate have. Of course, duplicates must be similar in some

---

[24] Having a duplicate may be a property of the duplicated object which it has in virtue of the duplicate. But it is not a property which it has in virtue of the duplicate's having the property of being a duplicate.

respects, because that is what makes a duplicate a duplicate. But whether duplicates are similar in respects of some property that is either intrinsic or extrinsic is a different matter. That my duplicate has my generic abilities does not prove that generic abilities must be intrinsic. For he (my duplicate) has those properties in virtue of me having it, and *I* might not have them *without* living in the actual environment in which I do.

The asymmetry of duplication-relation is particularly relevant when it comes to abilities, since one way to argue for – and not just to postulate in an ad hoc manner – the intrinsicness of abilities is to *presuppose symmetry* between the actual and the imagined impoverished situations. Suppose that being red is the ability (or disposition, the difference does not matter in the present context) of certain objects to elicit red sensation in us in certain standard circumstances. Can objects be red also in worlds in which no creature with the appropriate visual apparatus exists or where circumstances that count as 'standard' here never obtain?

If the answer to this question is negative, then colour-properties are obviously extrinsic. But notice that even if we answer *that* question positively, the property itself might be extrinsic. Suppose that being red is the property of eliciting *actually* a sort of sensation in us in certain standard circumstances. *That* property can be possessed by duplicates of some (actually) visible objects even in some impoverished circumstances – or 'contracted worlds' – in which there are not any creatures that can have red sensations.[25]

However, this can be true only because the duplication-relation is not symmetric. And this is important. For, from the perspective of the 'duplicate's world', there is no way to understand how anything *there* can be red (provided, of course, that redness is indeed the ability or disposition to elicit red sensations in us, etc.). Considered from that perspective, it seems to be an extrinsic property of objects that they are red; because they would have those properties only if some creatures with the capacity of having red sensations existed too.

Here is another example. Cyanide is a toxin *in our world* because it can disrupt a biochemical process essential for life. Thus, it has the ability to poison because there are organisms with a certain cellular structure the normal operation of which can be disrupted by cyanide. Should we say that cyanide is toxic in some impoverished circumstances in which organic matter does not evolve? If cyanide's ability to poison is interpreted as intrinsic, then we should. Cyanide

---

[25] Without using the notion of duplication, it is this kind of argument that John Heil seems to be relying on when he argues against the idea that powers are 'relational'; which means, if I understand him correctly, that they are extrinsic. See Heil (2003, 91–2).

has the ability to poison humans *actually*, and cyanide must have *that* property also in imagined nonactual circumstances in which no organic life evolves.

But now imagine a situation in which a kind of organism evolves for which cyanide is life-sustaining. In that situation, cyanide is nutritious rather than being poisonous. Is this sufficient for regarding cyanide to be nutritious in *our* world? Cyanide is *toxic to humans*; or, in general, to organisms that operate on the basis of the same kind of metabolism as humans do. Whether or not it is also *nutritious* depends on whether or not there are organisms for which consuming it is life-sustaining. But then, the possession of the relevant ability *must* be extrinsic in the sense that it is sensitive to which objects exist in a universe.

Returning finally again to my ability to drive cars, I do not deny that my 'lonely duplicate' has that ability. My point is that this does not show at all that this ability is not extrinsic to a certain extent. The possession of that property still depends on what else exists in the actual environment of its bearer; and, in this sense, that property is also extrinsic. Intuitions about duplications in imagined impoverished situations cannot help explaining what makes, if anything, some abilities intrinsic. We need an alternative approach.

The sense of sensitivity

I propose then to understand extrinsic abilities as properties the possession of which can be variably sensitive to factors outside its bearer *in a special way*. It is important to understand clearly what that special way is *not*.

First, in many philosophical contexts, sensitivity is understood causally. In the causal sense, an object is sensitive to its environment if the ways in which it changes and acts are causal responses to some changes in its environment. The sound level produced by the speakers in my room is sensitive in this sense to the changes of the states of the amplifier to which they are connected. But even if causal sensitivity has an obvious relation to the *exercise* of certain modal properties – being sensible in the causal sense is grounded in objects' dispositional properties after all – it is not the condition of their *possession*.

If we understood sensitivity in causal terms, then we would trivialize the claim that many modal properties are extrinsic. Since most objects' properties can change as a result of some extrinsic influence, sensitivity in the causal sense could not distinguish extrinsic abilities from intrinsic ones, or discriminate among the various degrees of extrinsicness, in the way in which I argue it does. An uncharged battery can be recharged and hence acquire some ability. But whatever the sense is in which this change indicates a property's sensitivity to its possessor's environment, this is not the sense in which being charged is extrinsic to a certain degree.

Second, 'extrinsic' is sometimes understood as being relational. However, both predicates and properties can be relational. Relational predicates need not express relational properties. Jack being 6 feet tall might be interpreted as a relation between him and a number. But this would hardly show that agents' specific heights are relational properties. Moreover, even if a relational predicate does indeed express a relation, relational properties need not be extrinsic. For instance, 'being a proper (material) part of' is a relational predicate that expresses a relational property, but, if anything, it is not an extrinsic one.

It is for this reason that, in the case of modal properties like abilities and dispositions, extrinsicness should be understood with reference to a sort of sensitivity to environmental factors which is neither causal nor simply relational. The question about extrinsic abilities concerns a special sense of sensitivity that can explain why the extrinsicness/intrinsicness of modal properties is a matter of degree.[26]

I shall begin to explain that sense of sensitivity by reinterpreting an often used, and indeed very illustrative, example about the ability of a key to open a lock.[27] The question is whether, and if yes how, a key's possession of that ability depends on the existence of the appropriate lock. It is initially clear that even if a key has never been used to unlock a door, it can still have the ability to open a certain kind of lock, if that kind of lock actually exists. This is the consequence of the distinction between conditions in which abilities are exercised and the conditions in which they can be possessed by an object. An object can have an ability even if it is never actually exercised.

But suppose that a key has been so badly manufactured that it cannot open any door. Suppose, further, that someone who, for some reason, insists on using it, designs and manufactures a lock to which it fits. It seems natural to say that, in such circumstances, the key *acquires* an ability to open some locks in virtue of a change in its actual environment. Imagine the more realistic inverse case. A key has the ability to open certain locks for a while, but, in one unfortunate moment, all the locks that it can open are destroyed. In those circumstances it

---

[26] In this, I agree with Barbara Vetter's view; see Vetter (2013).
[27] The example has been introduced by Richard Boyle in the seventeenth century, and then discussed more recently, among others, by Shoemaker (1980), Molnar (2003, 102–5), McKitrick (2003) and Vetter (2013).

sounds natural to claim that the object *lost* an ability. It has lost the ability to open any door.

What follows from all this? An object's ability to open some lock is a rather generic ability as compared to its ability to open one specific sort of lock; or the ability to open the lock on my desk's drawer. What this shows is that even rather generic abilities – as the ability to open some locks – can be possessed, acquired or lost in a world as a result of some change in the actual environment of their bearers. Thus, even very generic abilities might be extrinsic in the sense that their possession is sensitive to their bearer's environment.

Consider again a 'key-shaped' object without the corresponding lock. Suppose that the kind of lock which can be opened by the key has never existed and never will exist. That object, I would claim, lacks the ability to open a door. Denying this would lead to some absurd conclusions. For instance, we should say that *any object at all* does *actually* have the ability to open locks (and hence anything whatsoever is a key), since locks of all fancy kinds could be designed to which they fit. As a matter of fact, this is not just an imaginary example. Today, we usually receive magnetic cards in hotels to open our room's door. No such card would, if had existed, have had the ability to open a door hundred years ago. Magnetic cards have that ability now only because there are locks which have the ability to be opened by them.

Losing and acquiring abilities are certainly 'real changes', not 'mere Cambridge-changes'. They are changes of objects' properties, which can be essential to their operations and hence must be 'real'. And what *can* happen in the world after those changes, changes too: new possibilities arise, or old ones disappear.

When no corresponding lock which can be opened by a key exists, a key-shaped object loses its property not only to open this or that lock but to open *any*. The object is no more a key. An object can exemplify a sort of artefact only if it possesses certain generic abilities. Thus, losing an ability as a (metaphysical, not causal) consequence of some change in the object's environment might entail that an object ceases to exemplify the sortal it exemplified before. Why should we not consider such changes as 'genuine' (rather than a 'mere Cambridge-change', whatever that means)?

Recognizing that the question of extrinsicness is a matter of degree can help sorting out some intuitions about the extrinsicness of modal properties like abilities and dispositions. The ability of a key to open my front door is obviously extrinsic in the sense that whether an object possesses it depends on the ways in which some other particular object is. Is the ability of that key to open doors also extrinsic? This ability is more generic in the sense that its possession is *less* sensitive to changes in the actual environment. For instance, changing the distribution of locks on doors will not affect it. But if my arguments earlier are correct, it is not *entirely* insensitive to changes in the actual environment.

Think of an extrinsic property that does not seem to be modal (though it might be a manifestation of some abilities) like Jane's being married to Jones. If Jane divorces Jones and she marries someone else, one of her extrinsic properties changes. But she does not change with respect to her extrinsic property of being married (to someone). If James makes her a proposal at a wrong time, it is enough for her to answer: 'Sorry, I'm already married.' It does not matter who the lucky one is. However, that more generic property can change in the moment in which no one is Jane's husband.

Similarly, an ability to φ might be an extrinsic property if whether or not o has the property depends (logically or metaphysically) on whether *another particular* object o^* *with some particular feature F* actually exists. Even very specific abilities can be instantiated by more than one object: one can have more than just one key to her front door after all. But if the lock of *that door* has been changed, all of those objects lose their specific ability.

However, an ability can be extrinsic in the sense similar to 'being married (to someone)' is. A generic ability is extrinsic if its possession requires the existence of *some* object(s) *with some specific feature*. My key has the ability to open the kind of lock with which my desk's drawer is equipped. Changing the lock on my desk will not change that more generic ability of my key. But even its more generic ability is extrinsic in the sense that whether it is possessed by any object is sensitive to the existence of the kind of lock that can be opened by a kind of key. Thus, the extrinsicness of generic abilities is a matter of degree, and not an all-or-nothing matter.

Having the ability to open a door is, of course, not a 'natural ability' in the sense that it depends on human design and purposes. But my argument does not rest on any assumption that is specific to artefacts. Generic natural abilities – like being hard or fragile, soluble or insoluble, charged positively or negatively, – can be extrinsic *to a certain extent*.

There is nothing counterintuitive in the claim that some generic abilities which are 'natural' (in the sense of not being social, conventional, a product of intentional design, etc.) are extrinsic to a certain degree. Nothing could be fragile or sturdy in a universe in which temperature is so high that matter never solidifies. And nothing could be soluble in a universe in which nothing exists in a liquid state; cyanide would not be poisonous if living creatures had not evolved on Earth. Thus, even generic natural abilities can be extrinsic to some degree.

It is sometimes claimed that at least modal properties which are 'fundamental properties' must be intrinsic.[28] But it is unclear why modal properties' extrinsicness would 'fade away', as it were, just because they are (or are considered as) 'fundamental'. As we shall see in Chapter 5, the possession of *any* modal property that plays a role in the understanding of how objects can interact seems to be sensitive to *what other properties* are instantiated in the universe. Only those modal properties are purely intrinsic the exercise of which requires no interaction with the object's environment. But such modal properties are typically not 'fundamental'.

We need to distinguish at least two senses in which we can talk about the fundamentality of some properties. In one sense, a property is fundamental if anything that exists has it in virtue of being existent. If some version of physicalism is true, for instance, then some physical properties might be fundamental in this sense. In another sense, fundamental properties are properties of the fundamental constituents of the world, as fundamental particles are assumed to be. According to the first interpretation, electric charge is not a fundamental property given that there are uncharged physical objects even at the particle level; but it is fundamental in the latter sense.[29]

Let us assume – without being committed to this – that there is a fundamental (microscopic) level, and fundamental abilities and dispositions are modal properties that can be instantiated at that level.[30] Fragility, hardness, solubility and so on might be natural properties, but they are not fundamental in that sense. Then, the issue about intrinsicness concerns whether it makes sense to ascribe any of such 'fundamental abilities' in a *global* environment ('in a world') in which certain other abilities are not instantiated. And it does not seem to me that it does.

We ascribe abilities at the fundamental level, at least partly, in order to explain some macroscopic phenomena with reference to some physical interactions at the microscopic (particle) level. Hence one of the obvious conditions for microphysical entities to have certain abilities seems to be that other entities which are necessary for the relevant kind of interactions have the appropriate abilities as well. A lonely particle existing in an imagined impoverished 'world'

---

[28] Such a view has been defended, for instance, by Bird (2007).
[29] For instance, different types of neutrinos are uncharged fundamental particles. For more on the problem of fundamentality in this sense see Huoranszki (2019).
[30] For arguments against the view that there must be a fundamental level, see Shaffer (2003). For a different account of fundamentality in physics see, again, Huoranszki (2019).

cannot have the abilities and dispositions which 'fundamental particles' have in our rich and 'multilevel' universe.[31]

To sum, abilities ground possibilities, and what possibilities there are depend in various ways on how *the world as a whole* is. The more specific an ability is, the more sensitive its possession is to the changes of the circumstances of their bearers. But the possession of many generic abilities, particularly those that are manifested by objects' interactions, can be sensitive *to some extent* to the actual environment. Many generic abilities do not 'freely recombine' with other properties; their instantiation entails that other properties are instantiated in the same universe. And, as we shall see, it is their instantiation in the same universe that determines which kinds of events are possible in that universe.

Intrinsicness and systems

When the question about the extrinsicness of modal properties arises, philosophers typically have in mind individual objects, ordinary substances like glasses (which can break), sugar cubes (which can dissolve) or cats (which can chase mice). Harré and Madden, for instance, says that the ontology of powers is the ontology of 'powerful particulars'. Such particulars may not be ordinary objects. They might persist only for a fraction of a second, they might be fields rather than material objects or they might be only 'virtual particles'.[32] But they are particulars in the sense of being numerically distinguishable individuals.

So far, in discussing specificity and extrinsicness, I followed this tradition. My main question was whether or not abilities are intrinsic properties of objects or some samples of chemical substances. Nonetheless, it is crucial to note that not only particulars but also ensembles of particulars can have modal properties. We may call such ensembles systems.

Dispositions and abilities can indeed be intrinsic to a system. However, a system's intrinsic abilities (or dispositions) are the results of the extrinsic (to

---

[31] Molnar claims that all extrinsic dispositions of objects apply to them in virtue of their intrinsic properties as it is shown by physics (Molnar 2003, 108–9). I do not think that physics has shown anything like this. Consider some philosophers' favourite example of 'simples': quarks. Their role is, by definition, to explain the plethora of baryons which have been identified by nuclear physicists; hence quarks' properties are identified with the purpose to satisfy that need. Quarks' properties simply couldn't exist without interacting with others thereby constituting a larger particle.

[32] However, it is important to note that events cannot be the bearers of modal properties. Philosophers sometimes talk about the 'power of an event' to bring about another, but this strikes me as a confusion. Objects might have the power to bring about an event, and the exercise of that power might itself be an event. But we should not ascribe powers to events. I say more on how events and modal properties are connected in Chapter 6.

certain degree) abilities of their constituents. Systems, in the sense I understand them here, consist of *interacting* particulars. Thus, abilities that are intrinsic to a system are grounded in the abilities of its constituting elements that interact in certain ways. And those abilities are extrinsic exactly in the sense, and to the degree, in which the key's ability to open some doors are extrinsic.

This means, contrary to what is often assumed, that intrinsic modal properties are grounded in the extrinsic ones; or more precisely, the *more* intrinsic abilities and dispositions are grounded in the *less* intrinsic ones. This might sound paradoxical. As we have seen, the standard assumption is that intrinsic properties (e.g. spatial extension like height) ground the extrinsic ones (Plato's being taller than Socrates). Nonetheless, in this final section, I am going to argue that there is nothing paradoxical about this.

There are many things which cannot be done by individual atoms (or, more generally, individual particles) in our world, but which their ensemble can do. Ensembles of particulars (objects or particles) which constitute systems are especially important in physics. A certain volume of gas in a container has the ability to increase the pressure on the container's wall when heated. A certain volume of gas is not a particular object: it is an ensemble of particles (atoms or molecules). Water in the sea is not a particular object; but it is able to dissolve salt.

Matter in gaseous and liquid state is not an 'object' in any ordinary sense of that word. Clouds and oceans are not objects. But if current physics is right, then they consist of an ensemble of particulars. Such ensembles have abilities and dispositions that their constituents cannot have. An individual atom will not exercise pressure on the wall of a container, and it will not dissolve anything.

The ability or disposition of a volume of gas to exert certain pressure on the walls of its container is an intrinsic property in the sense that its possession is not sensitive to what happens outside the system or how it interacts with systems. However, its constituent atoms' abilities are extrinsic to a relatively high degree. For how those atoms *can* behave depends on how they can interact with each other.

Although it is perhaps less obvious, the same is true about solid objects. What we normally consider as an object's intrinsic ability or disposition is determined to a large extent by which abilities its constituents have. Coal is malleable – and can be easily scratched – while diamond is especially hard; that is to say, it has the ability to resist being deformed or scratched. Their different modal properties can be explained with reference to their constituents' abilities and dispositions to stably arrange in different structures. But those abilities are extrinsic to a certain degree, for, ultimately, they depend on which kind of interactions their bearers can participate in.

The fuel in a nuclear power station equipped with an appropriate safety device has the generic ability to explode when temperature rises to a certain level.[33] But it does not have the *specific* ability to explode in the type of circumstances in which it is actually found, and where the temperature, thanks to the presence of some safety device, cannot reach that level. What about nuclear power stations themselves? Such stations are physical *systems* that contain both the fuel and the safety device as their parts. Thus, power stations (if we are lucky enough) do not have the specific ability to explode; neither are they disposed to do so.

Philosophers sometimes call certain dispositions 'partners' to each other. Such partners work reciprocally.[34] It is characteristic of such dispositions that their manifestation requires the presence of (at least) two objects with 'matching' abilities or dispositions. Typical examples include sensible properties, like the disposition to elicit red sensation in certain observers in standard conditions and the disposition that a red sensation is elicited in similar circumstances. The object's and its perceiver's dispositions mutually depend on each other in the sense that they can be manifested only together.

There is no reason, however, to restrict the occurrence of such partnership to sensible properties or to the so-called secondary qualities. Nothing can be a solvent without their being some appropriate soluble; and nothing can be scratched without something else being hard enough to scratch it; nothing can be fertilized without something having the ability to fertilize it, and so on. Most abilities and dispositions the possession of which governs objects' interactions seem to come in pairs (or in trios, quartets, etc.).

In fact, the idea of 'disposition partners' is just a rediscovery of the Aristotelian idea of 'active' and 'passive' powers. We can retain the essence of the Aristotelian idea, without following Aristotle in distinguishing the 'active' and the 'passive' part, which is motivated by the particular examples Aristotle often relies on (like the example of a teacher actively passing knowledge to his student who 'receives' it).[35]

---

[33] This is an example from Mellor (1974).
[34] See, for instance, Lewis (1997/1999), Martin's contribution to Armstrong, Martin and Place (1996), and Heil (2003, 83). I shall return to the problem of interacting abilities and their relevance to the analysis of modal properties in more detail in Chapter 5.
[35] See Aristotle *Physics*, Book III, 3. Most contemporary philosophers who follow the interaction-model reject Aristotle's distinction. Martin (2008) is mainly responsible for resuscitating Aristotle's idea of reciprocal capacities, but he rejects Aristotle's distinction between active and passive powers. For further developments, see Heil (2017) and Anjum and Mumford (2018). For an interesting recent defence of the distinction between active and passive powers, see Marmodoro (2017).

My coffee has the ability to dissolve sugar, and the sugar has the ability to sweeten my coffee. Which one is active, and which one is passive? It seems that it depends on our envisioned aims (dissolving the sugar or sweetening the coffee), and not on the nature of the modal properties themselves. What does seem to be true, however, irrespective of how we describe the process, is that one of these substances cannot have the ability without there being some other substance with the corresponding abilities.

But the instantiation of certain modal properties can also depend on the existence of other objects with certain appropriate abilities without any 'partnership'. For an obvious example, think of those chemical substances' abilities which we call catalysts. Catalysts do not themselves participate in the chemical reactions of other substances, but they have the ability to increase the rate of some such reactions.

In some cases, like in the case of biocatalysts like enzymes, their presence is necessary for the 'partnership' of certain other abilities and dispositions. It is a generic feature of dairy products that they have the ability (or even the disposition) to nourish most humans. However, as the phenomenon of lactose intolerance testifies, having that ability depends on the presence of a kind of enzyme (lactase) with a certain ability in the consumer's organism. The ability to nourish must have 'a partner': an organism with the appropriate kind of enzyme. But even if dairy products' ability to nourish depends on the presence or absence of that enzyme with the appropriate ability, enzymes are not 'disposition partners'; since, obviously, it is not the enzyme that nourishes us.

These examples are well known. Abilities or dispositions which can be exercised only by objects' interactions – never mind whether the interaction requires partnership or not – require the existence of distinct objects or substances with some corresponding abilities. Nevertheless, many philosophers would deny that the relevant modal properties are extrinsic to any degree. Their reason seems to be that even if we ascribe such properties in order to understand how things can interact, and this makes it unavoidable that we mention other objects and properties in order to specify the ability, this does not entail that the property so specified itself is extrinsic.

However, in the light of our earlier discussions, it is unclear why this should be so. Of course, the fact that certain properties can be *exercised* only in certain conditions does not make modal properties extrinsic. But in the case of interacting objects' abilities, it is not the ability's exercise, but its *possession*, which seems to depend on some features of the object's global environment. As I have argued, it simply does not make sense to ascribe certain modal properties without some other's being instantiated in the world.[36] Many very generic modal properties are abilities to participate in some sort of interactions, so that particulars cannot possess those properties without having – 'in the same world' – 'partner objects' with the corresponding abilities.

Importantly, as I mentioned in the previous chapter, this does not mean that there *must* be an object which has the more specific abilities of which the more generic one is only a determinable. Suppose that some particle of type P has the ability to participate in an interaction of type *I* when it meets particle type Q in circumstances C. If circumstances C never obtain when a particle P and a particle Q meet, then P never has the *specific* ability to participate in interaction *I*. Nonetheless, it can have the more abstract ability to participate in such type of interactions.[37]

There might be types of interactions in which objects, given their abilities, *can* participate, but they never actually do. Even if no fuel in the nuclear power stations had ever exploded *because* of the efficiency of precautionary devices, and hence none had had the specific (in the circumstances) ability to explode, they would still have the more abstract ability to explode in certain unactualized conditions; for instance, in conditions in which the precautionary devices are absent or malfunctioning.

I must admit that, as with almost any philosophically interesting distinction, there is certain vagueness about whether we should regard the bearer of an ability a system or a particular. Temperature is an essentially macroscopic phenomenon, and as such it is always ascribed to some system of particles. But our bodies are particular macroscopic objects, which have the ability to keep a constant body temperature. This ability is then ascribed to an individual organism, and not the 'mere ensemble' of particles.

Steam in a container or water in a cup is not normally considered as particular objects, but rather as ensembles of particles or 'mere stuff'. Nonetheless, when they are cooled down, and hence freeze and solidify, they can become one. Solid macroscopic bodies are archetypes of particular objects to which we ascribe

---

[36] Admitting this difficulty with the intrinsicness-thesis, Williams suggests 'power holism' as a solution. However, it is very hard to see how holism fails to imply that modal properties are extrinsic in some sense. Holism is supposed to be the thesis that what certain powers are depends on all the others as 'each contains within it a blueprint of the entire universe' (Williams 2010, 95). Thus, if one power is missing, it changes all the rest. How is it possible then that powers are not extrinsic to any degree? How can such a blueprint *of* other powers be intrinsic? Williams seems to hold it so because 'power holism' does not entail that all the powers in the blueprint must be instantiated in the same universe. But, as I have already argued in the previous section, this is conceivable only from the perspective of a universe in which they *are* actually co-instantiated.

[37] This is Tooley's famous example for the non-supervenience of causal laws on particular events; see Tooley (1997).

abilities and dispositions, even if they are constituted by the same ensemble of microphysical particles which constitute a volume of liquid or gas.

However, there are clear enough cases when it would be bizarre to assume that the bearer of a modal property is a particular. The Sun is disposed to produce solar flares; this is a disposition of a particular (to the extent that the Sun, which is a volume of gas, is a particular, which might be a further issue). The Solar System is disposed to be shielded from cosmic rays; this is a disposition of a system. The gas in the container has the disposition to exert a certain pressure on the wall of the container; this is a disposition of a system. The container is disposed to blow up when the temperature reaches a certain level. This is a disposition of a particular.

Individual animals have certain abilities that ground their disposition to survive in a certain environment. But biological species too, considered as an ensemble of individuals, might have or lack the ability to survive. The bearers of such abilities and dispositions are not particulars, but either *species* of objects (individual organisms or types of genes, etc.) as in the case of biology, or *ensembles* of particulars, as in the case of a closed physical system. System's dispositions can also play a role in social sciences. An economic system which is in a certain macro-economic state is disposed to have certain rate of inflation if certain policies are adopted. But an economy is not a particular; it is an ensemble of people who participate in countless specific economic transactions.

Finally, there are also statistical dispositions that can be ascribed only to species or ensemble of objects, and not to individuals. Importantly, *statistical dispositions* should not be confused with *the statistical interpretation of dispositions* which I proposed in the previous chapter and shall discuss further in the next two chapters. It might be true that the average member of a certain population is disposed to have three and a half offspring. This does not entail, of course, that any individual member of that population is disposed to have three and a half offspring. No individual can have that ability. Statistical dispositions – important for many sciences – can be ascribed only to some ensembles of particulars, and not to their particular constituents.

As a consequence, it is important to distinguish systems, understood as certain ensembles of objects, and particular objects themselves, as possible bearers of modal properties.[38] Most physical abilities and dispositions which are intrinsic – or intrinsic to a relatively high degree – are intrinsic dispositions of particular systems, not of particular objects. And most, perhaps all, modal properties of their constituent elements are extrinsic to a variable degree. If they were not, they could hardly ever constitute a system.

---

[38] About the importance of physical systems as bearers of dispositions, see further Hüttemann (1998).

4

Modal properties and conditionals
The nature of the connection

In the previous chapter, I have argued that objects can possess more or less specific abilities which ground more or less abstract possibilities. The purpose of this chapter is to argue that we can identify abilities at various levels of specificity through their logical connections to counterfactual conditionals.

For a long while, most philosophers thought that the ascription of modal properties like dispositions and abilities is logically or conceptually linked to the truth of certain conditionals. The so-called conditional analysis was meant to uncover the logical structure of this connection. Many contemporary accounts of modal properties reject the conditional analysis. The main aim of this – and a partial aim of the next two – chapter is to *reinterpret* and defend it.

Most debates over the conditional analysis centre around some technical difficulties concerning the ascription of modal properties. While I cannot, and do not mean to, dismiss such difficulties, I am more interested in the changing philosophical and metaphysical presuppositions which motivate these debates. I shall argue that we have a distinct theoretical motive for trying to defend some version of the conditional analysis. If the ascription of abilities does indeed entail the truth of some conditionals, then first-order modal properties, rather than possible worlds, can be taken to be their truth-makers. And if conditionals entail the ascription of abilities, then we can use the former to specify, *at various levels of generality*, the content of the latter.

The debates in which I will engage in this chapter and the next have traditionally been about dispositions. I think that, to a large extent, the arguments used in these debates apply in the context of abilities as well. Both the ascription of abilities and the ascription of dispositions are conceptually connected to the truth of certain conditionals. However, the counterfactual conditionals to which these two different types of modal properties are connected are different. The reason for this is, again, not merely semantic, but is related to the different ontological roles of the two different types of modal properties. By connecting the ascriptions of abilities to counterfactual conditionals we identify contingent possibilities; by connecting the ascription of dispositions to counterfactual conditionals we identify nomic and teleological regularities or habitual tendencies.

I shall begin the discussion by focusing on the traditional objections against the conditional analysis, which concern dispositions. Next, I shall investigate the relevance of the objections in the context of abilities. Then, I shall suggest a minor modification of the traditional analysis which, I shall argue, can answer the standard objections against the analysis. Finally, I shall offer a new type of conditional analysis of nomic and teleological dispositions.

The purpose of a conditional analysis: Phenomenalism, ontological reduction or enhanced understanding

As mentioned in Chapter 2, logical empiricists, the philosophers who first discussed the problem of dispositions, were sceptical about the very idea of metaphysics. They were interested in philosophical semantics. Their aim was to provide a semantic analysis of the meaning of sentences containing 'disposition terms'. They thought that an analysis was needed because the meaningfulness of this type of statements constituted a challenge to their verificationist program. On the one hand, such statements are often used in many advanced sciences so they must have semantic content (unlike the assumedly unverifiable and hence meaningless statements about metaphysics). On the other hand, their truth cannot be directly verified by observation. Hence the purpose of the positivists' program: they thought we need to explain how statements containing disposition terms are verifiable by providing a logical link between such statements and what is observable and hence empirically testable.

It is important to keep in mind that the *original theoretical motive* for a conditional analysis of dispositions was the commitment of early-twentieth-century empiricists to verificationism and phenomenalism. Their project was to explain the meaning of statements containing disposition terms by 'translating' them into statements about observables: that is to say, statements that report literally the *manifestations* of dispositions. Manifestations are understood here as events which are, or at least can be, directly observed. However, since observ*ability* is just another disposition, the only *genuinely* non-dispositional features of the world were taken to be *events*, understood as *singular occurrences of some specific type of observation.*

However, it was clear from the beginning that a single (type of) observation-event cannot be sufficient for specifying the meaning of statements containing disposition terms, because distinct dispositions might be associated with the same kind of observed changes. Changes of temperature, colour or state of motion might manifest entirely different dispositions of objects. Hence, in order to distinguish different disposition terms, we also need to make some reference to the wider observable circumstances in which these perceptible changes occur.

Thus, originally, the conditional analysis was formulated as an attempt to capture the proper connection between manifestation-events and the circumstances in which they occur in order to specify the meaning of a disposition term. For this purpose, a mere statistical association between circumstances and manifestations cannot be sufficient. For even if it is true that an object (or a kind of objects) has always been observed to φ in the local laboratory, we would not normally ascribe to objects or substances the disposition to φ on that ground. In order to define the meaning of a disposition term, we need to assume that the circumstances of manifestation play some role in the object's manifesting the disposition at a time. Put briefly, the occurrence of the manifestation must be a function of the circumstances. Conditionals were meant to capture this functional connection.

In the early debates over the meaning of disposition terms, several attempts have been made to understand the functional connection between manifestations and its circumstances with the help of material conditionals. These attempts were motivated by the assumption that the logic which is meant to represent the formal structure of a language, or at least the language of sciences, must be extensional. This attempt failed for well-known reasons.[1] It has been become clear that functional dependency is a modal connection, and hence we need to use some 'stricter' conditional in the analysis of disposition terms.

A further important characteristic of the early approach to the problem of dispositions was that the phenomenalists entirely neglected the role of *objects* and dispositions *as their properties* in the account of disposition *terms.*[2] Critics of phenomenalism rightly complained that we cannot understand the nature of dispositions unless we assign a proper role to objects and to their properties

---

[1] For brief surveys of the attempts to provide an extensional analysis and of the reasons why they all failed see, among others, Mellor (1974) and Mumford (1998), Chapter 3.
[2] In fact, as we have seen in the previous chapter, the very notion of a persisting object as the bearer of such properties was understood as dispositional by some empiricists.

in our analysis. But this does not settle the issue about the conditional analysis either. For there are two fundamentally different views about the nature of the properties that may be relevant to the analysis.

On one account, the relevant properties cannot be modal properties. Since objects' nonmodal properties are often – though as we have seen in Chapter 1, somewhat misleadingly – called 'categorical', we might call this the 'categorialist' view.[3] According to this view, objects can have a disposition only if they possess some nonmodal, categorical property or properties which serve as the 'bases' or 'grounds' of their dispositions. What role these properties might play in certain attempts to provide a conditional analysis of modal properties I shall discuss later in the section titled 'Losing and gaining abilities'.

On an alternative view, the ascription of modal properties is not logically or conceptually connected to the use of subjunctive conditionals at all. Dispositions are first-order modal properties of objects, which we can identify simply with reference to the kinds of events that are their manifestations. This is the view that I called in Chapter 1 the 'intentionalist' approach to modal properties.

In this chapter and the next, I shall argue for a third view. I grant to the intentionalists that modal properties like dispositions and abilities are first-order properties of objects (or substances, or ensembles of objects). But this does not mean that the ascription of such properties is not logically connected to the truth of some conditionals. I have already briefly indicated why I reject intentionalism about modal properties in Chapter 1. I shall come back to this issue in more details again in Chapters 5 and 6.

Some philosophers think that the ontology of modal properties should be entirely independent of the semantic analysis of disposition terms. However, as the debates over the role of modal properties in the analysis of dispositions demonstrate, semantics cannot be entirely free of ontological commitments. The ontological and semantic *aspects* of the debate are, no doubt, distinct. But we cannot argue for or against some version of the conditional analysis as a *philosophical* analysis of modal properties without some commitment about what such properties really are. The issue about the connection between disposition-ascriptions and the truth of counterfactual conditionals is not *merely* semantic. Any possible view about that connection shall presuppose some views about the role of modal properties in ontology.

---

[3] As I have explained in Chapter 1, I find this terminology misleading. I shall say more on the idea of such nonmodal properties later in Chapter 7.

I shall argue for a version of the conditional analysis of modal properties which assumes that dispositions and abilities are first-order modal properties. Objects possess such properties exactly in the same sense in which they possess their nonmodal properties (if indeed there are *any* such properties; more on this in Chapter 7). Even if it is true that, historically, the conditional analysis was motivated by phenomenalism and then by some kind of reductionism (of the modal to the nonmodal), we can reject both, and nonetheless accept the philosophical importance of a conceptual connection between ability-ascriptions and the truth of certain subjunctive conditionals.

Further, it is important to note that, as far as the ascription of modal properties is concerned, there is nothing special about disposition *terms* as such. For, as we have seen in Chapter 2, there are various different ways in which we can ascribe modal properties. 'Jill can read French' ascribes an ability to her. Dispositions can be expressed even more simply without the use of any modal auxiliary. 'Jack collects stamps' ascribes a dispositional property to Jack (meaning that he is a stamp collector). How we ascribe modal properties may also depend on which language we use. It is not a matter of *terms*.[4]

But if we reject both phenomenalism and reduction, why should we be interested in the logical connection between modal properties and conditionals? Because, first of all, counterfactual conditionals play an important role in both our theoretical and our practical reasoning. Antirealists about modal properties might claim that such reasoning can be best understood in terms of possible worlds. In contrast, according to the realist view, the ultimate ground of counterfactual reasoning is to capture the modal structure of the actual world. And the best way for us to capture the modal structure of the world is to connect the ascription of modal properties in it to the ways in which we use some subjunctive conditionals.

Moreover, with the help of the conditional analysis, we can further elucidate the difference between certain types of modal properties. For, as I shall argue, abilities and dispositions are connected to different types of subjective conditionals. Finally, and most importantly, we can determine the degree of specificity of modal properties (of either type) only through their connections to the truth of certain conditionals. In general, we can learn much about the nature of contingent possibilities as well as of nomological connections if we

---

[4] In German, for instance, 'Es lässt sich sehen' can mean, more or less, that 'it can be seen' or sometimes even 'it is visible'.

understand better how the ascription of abilities is connected to counterfactual conditionals.

The traditional conditional analysis of dispositions and the standard objections

In this chapter I shall reinterpret, and to a certain extent resuscitate, the traditional conditional analysis, or TCA for short. As I understand it, TCA offers a general schema for specifying the content of modal properties; most importantly for our purposes, a schema for specifying abilities. In this section, I shall summarize the most important objections against TCA as they were originally raised. These objections concerned not so much, if at all, the conditional analysis of abilities, but the conditional analysis of dispositions. However, I shall argue that the *philosophical import* of the counterexamples raised against TCA are largely – though, as we shall see, not entirely – similar in the two contexts. It is another matter that, given the already explained differences between the respective types of modal properties, the objections require somewhat different responses in the two contexts.

There is no universally accepted standard formulation of TCA. But the following schema will serve as a good starting point since, with reference to it, I can well summarize the main objections to the idea that there is a conceptual connection between the ascription of modal properties and the truth of certain counterfactual conditionals.

> (TCA) For any object (or substance, or ensemble of objects) o and time t, there is a time $t' \geq t$ such that: o has the disposition D at t if and only if an event of type Φ involving o would occur at t', if o were in condition C at t.

When this schema is applied, we first substitute letter 'D' with some specific disposition (like fragility, solubility, electric conductivity) on the left-hand side of the biconditional; and then we 'analyse' it by substituting letters 'Φ' and 'C', respectively, with a description of some type of event and some type of conditions on the right-hand side.

The arguments against TCA are based on some counterexamples which challenge the truth of this biconditional. Using counterexamples against some analysis is, of course, an important tool in doing philosophy. However, counterexamples are only the beginning, and not the end of a story. The further and deeper issue is whether we should *abandon* or *modify* an analysis as a response to these examples. In this chapter, I shall argue that the counterexamples to TCA do indeed reveal some interesting philosophical issues about the nature of modal properties, but they do not justify the rejection of all possible versions of the conditional analysis.

For this reason, I am not going to present the objections in the order in which they appeared in the history of the debates over TCA. I shall present them in an order which facilitates the exposition of my own views concerning how TCA should be modified in order to provide an adequate account of the content of modal properties.

Masks and antidotes.[5] In the previous chapter I mentioned the case in which arsenic is taken together with an antidote. Arsenic is a toxin. It is disposed to, and able to, poison those who ingest a certain amount of it. But it does not poison those who ingest it and take antidotes.[6] Thus, it seems false that if arsenic were ingested at t, then those who ingest it would die before t'. Some people do not.

Similarly, I also mentioned the case of sodium and common salt. Sodium is a corrosive material. It is disposed to, and has the ability to, burn living tissue with which it comes into contact. Nonetheless, when sodium is bonded with chloride as it is in common salt, it is not disposed to damage our body. So, it is not true that sodium would damage one's body tissue if it came to contact with it. We may say that the corrosiveness of sodium is *masked* when it is bonded in common salt.

Further, it seems possible that even a sturdy object can fall in such unfortunate ways that it (unexpectedly) breaks. Thus, it is not true that if it were dropped at t, it would resist splintering. Conversely, a fragile object can land on the hard ground in such a fortunate way that it does not break. Thus, it is not true that if a fragile object fell on a hard ground, then it would break. As it has been mentioned, objects might have an 'Achilles heel' and hence some of their actions can fail to manifest a disposition which they otherwise possess.[7]

In the case of masks, the initial circumstances in which a process occurs makes it impossible that the object manifest its disposition. In the case of antidotes, after the initiation of a process, another is launched which thwarts the manifestation of an object's disposition. In the third case, it is some peculiar

---

[5] The terminology derives, respectively, from Johnston (1992) and Bird (1998).
[6] I shall assume here that a subjunctive conditional can be true even if its antecedent is factually true; actual truth can be a supposition. This is an assumption that almost all current accounts of counterfactual conditionals share, although it was denied by some earlier accounts of counterfactuals; see, for instance, Mackie (1973, 70).
[7] See Manley and Wasserman (2008).

feature of the object itself which explains why a certain event happens (or, in other cases, fail to happen) in the circumstances in which it should if the TCA were correct.

We shall see the relevance of these differences later. Nonetheless, what all these examples seem to show is that the truth of a counterfactual conditional in TCA is not *necessary* for the ascription of some modal property. The truth of a conditional does not seem to be entailed by the ascription of the corresponding disposition.

Finks and reverse finks. Perhaps the most influential objection to TCA is based on the following type of cases. Suppose that something is water-soluble at t_0 if and only if it would dissolve before t_1 if it were immersed in water. But imagine that someone has the magical power to turn an otherwise insoluble object soluble whenever it is about to be immersed in water. Then it is true that the object would dissolve if it were immersed in water. Nevertheless, the object, by assumption, is insoluble (Goldman 1970, 199–200).

Most philosophers thought that such examples speak more against the possibility of magic than against the conditional analysis of dispositions. However, in a highly influential article, Charles Martin describes a case which is logically analogous to the magic-examples but does not involve the use of supernatural powers. An electro-fink is a device which can make a wire live when it is touched by a conductor. According to TCA, a wire is live at t if and only if electric current would flow through it, if it were touched by a conductor. But if an electro-fink is attached to the wire, the conditional is true even if the wire is dead (Martin 1994).

Electro-finks can work in the opposite way as well. It is possible that a wire is live, but, thanks to the presence of a 'reverse-fink', no electric current would run through it even if it were touched by a conductor. In this case, we can correctly ascribe a disposition to an object even if the corresponding conditional is false. Thus, the possibility of finks and 'reverse-finks' seem to show both that a modal property can be ascribed to an object even if the corresponding conditional is false; and that the conditional in TCA can be true, even if an object fails to have the relevant modal property.

Mimics. Consider a type of event, like the splintering of a macroscopic solid object. This is the observable manifestation of fragility. Sturdiness is the opposite of fragility: the ability to preserve objects' integrity even when they are exposed to some external force exerted on them; for instance, when they are struck. However, even a sturdy object can splinter in circumstances in which striking it generates some process that eventually results in its splintering (Smith 1977).

This shows that observable events like sudden disintegration might only 'mimic' a modal property like fragility. The occurrence of mimicking is supposed to show, as the presence of finks does, that the truth of the conditional in TCA does not entail that an object must have the relevant disposition.[8]

It is interesting to note that, in some cases, masking a disposition means mimicking another. When an object's fragility is masked, it mimics sturdiness. And when it only mimics sturdiness, its fragility is masked.[9] Masking and mimicking are obviously connected when dispositions are gradable on a scale. When sturdiness is mimicked, fragility is masked, and vice versa, because nothing is 'maximally hard' in the sense that it cannot break in any conceivable circumstances. Nonetheless, not every modal property is gradable in that sense, and hence mimicking one property need not involve masking another. And, as we shall see, mimics and masks raise different *philosophical* issues about the nature of modal properties.

Nondeterminate processes. Even if there is a specific condition in which a disposition would become manifest, the occurrence of that condition is compatible with the non-occurrence of the type of event which specifies the disposition, because the disposition is not deterministic. A fair coin is *supposed to have* the nondeterministic disposition to land heads up when tossed, but it is also disposed nondeterministically to land tails up when tossed. Whether the coin *does indeed have* these dispositions at all is, of course, a contentious issue. I shall argue in the next chapter that it does not. But if we want to answer an objection, we need to state it first.

'Unconditional' modal properties. Finally, it seems that the exercise of certain abilities or the manifestation of certain dispositions does not depend at all on which circumstances an object is. Perhaps, a piece of flexed tempered steel is disposed to break at a certain moment, but no specific type of circumstances can be identified in which this happens. Even if the probability of its breaking increases with its age, there is no identifiable condition in which it breaks exactly at the time when it does.

Table 4.1 summarizes the consequences of the various types of counterexamples to CTA.

It is natural to divide these objections into two distinct groups. One group of objections aims to prove that the truth of a conditional cannot justify the ascription of a modal property; whereas another group of objections aims

---

[8] We owe the term 'mimicking' to Mark Johnston (1992). The problem was first noted by Smith (1977).
[9] As it has been observed by Fara (2005).

Table 4.1 Summary of the objections to TCA

	The sentence which ascribes a modal property to o is:	The counterfactual conditional about o's behaviour when it is set in condition C is:
Masks, antidotes, Achilles heels	true	false
Reverse finks	true	false
Finks	false	true
Mimics	false	true
Nondeterministic processes	true	neither true nor false
'Unconditional' modal properties	true	uninterpretable

to prove that an object can have a modal property even if the corresponding conditional of TCA is false. The two groups of objections do not have the same philosophical significance. Their significance depends on which purpose the conditional analysis of modal properties aims to satisfy.

If our purpose is to show that the ascription of a modal property involves nothing else than a commitment about the truth of some conditional, then the first group of objections seems weightier. For if those objections were sound, then the truth of a conditional can never justify the ascription of modal properties and hence the aim with which we initially introduced TCA could not be satisfied.

Suppose, however, that we take modal properties to be real, first-order properties of objects. Then, it matters less whether or not the respective counterfactual conditionals entail the ascription of abilities. It might seem that our aim with the analysis might be satisfied, if at least some conditionals are entailed by objects' having certain modal properties; or even more leniently, if there is at least some loose, but systematic, association between the two.

In my view, however, even if we assign a prominent role to modal properties in our ontology, we still need to posit some tighter conceptual connection between their ascription and the truth of certain conditionals. For, if the ascription of an ability were not entailed by the truth of a counterfactual conditional, then we would be completely unable to *specify* the content of modal properties. And if the ascription of modal properties did not entail the truth of certain conditionals, then the success of our practical and theoretical reasoning could not depend on which modal properties objects in fact have.

I shall argue, however, that none of the counterexamples proves that there is not some pertinent connection between the ascription of a modal property and some counterfactual conditional. The rest of this chapter shall discuss masks, antidotes and Achilles heels. I shall return to the issue of nondeterministic abilities and 'unconditional' modal properties in the next chapter. The possibility of mimicking raises some deep metaphysical problems about the concept of manifestation and the nature of events. I shall discuss these issues in Chapter 6.

Masks, antidotes and circumstances

Masks and antidotes are counterexamples to TCA only if we presuppose that the circumstances in which a disposition would become manifest or an ability would be exercised can be specified by a single factor, often called the 'stimulus' of the manifestation. The word 'stimulus' clearly indicates that the circumstances are understood here as the *causes* of the manifestation event.

Suppose we want to specify an object's ability with the help of the following conditional: this object is such that it would break if it were dropped. By using this conditional, we ascribe an ability (or disposition like fragility) to an object; but we also seem to describe a type of causal process. Similarly, if we say that a snake is such that it would poison those who were bitten by it, we do not only ascribe an ability (or the disposition of being venomous) to it; we also describe a type of causal process in which snakebites cause poisoning.

If we were interested in abilities only because they can specify some potential causal processes, it might be important to identify *a* single type of event as a cause. That an event can be a cause only in specific circumstances does not make the causal counterfactual 'if *s* happened, *s* would cause *e*' false.[10] But we are not interested here in causation; our concern is the connection between the ascription of certain modal properties and the possibilities that they ground. And, in that context, we cannot always simply assume that the circumstances in which *a* 'stimulus' occurs are sufficient to exercise an ability.

However, as we have seen in the previous chapter, abilities can be more or less generic, and how generic they are is often determined by how (in)sensitive their possession is to various actual conditions which are often (though not always) are extrinsic to their bearers. It is only maximally specific abilities that can be identified by mentioning one single condition of their exercise since at

---

[10] As is well known, the connection between causation and counterfactuals breaks down in the cases of causal overdetermination, but this is a problem which need not detain us here.

ascribing them we already tacitly assume that all other conditions necessary for their exercise are satisfied.

Masks and antidotes are defined as types of circumstances which can thwart the exercise of an ability. Hence, when they are present, an object cannot have the maximally specific ability, specified by one single condition of its exercise. But this can be adequately represented by a correct interpretation of the conditional analysis.

As we have seen, the original purpose of what later became called the 'conditional analysis of dispositions' was to introduce the use of disposition terms into the language of science rather than explaining the meaning of some words in ordinary language. Similarly, I suggest that the purpose of what we might call the conditional analysis of abilities is to specify abilities *at different levels of generality* by applying the following general schema:

(CA^A) For any object (or substance, or ensemble of objects) o and time t, there is a time $t' \geq t$ such that: o has the ability A at t if and only if an event of type Φ involving o would occur at t', if o were in conditions C at t.

In order to obtain an instance of this general schema, we first choose to substitute letter 'Φ' with a description of a type of event and letter 'C' with a description of a set of conditions. This choice is guided by which, and how generic, ability we aim to specify when we substitute letter 'A' with a suitable term or description.[11]

The abilities we aim to identify can be more or less generic according to two dimensions. Often, we express the degree of generality of properties by using different predicates to describe objects. We say that something is navy blue, or that it is blue; or simply that it is coloured. We can also use numerical expressions to ascribe more or less specific properties to objects, for instance, we can say that a physical body is less than 10 kilograms or that it is exactly 9 kilograms. Similarly, the ability to speak a language is less specific, than the ability to speak Spanish. The ability to swim is less specific than the ability to butterfly. The ability to butterfly is less specific than the ability to butterfly a hundred metres within a minute. And so on.

In this sense then, some substance, objects or persons which have a more abstract, generic ability at a certain time t, might fail to have the more specific one. I can have the generic ability to speak a language, without having the more specific one to speak Spanish. I can have the ability to butterfly, without having the ability to butterfly a hundred metres within a minute. The level of generality

---

[11] Thanks to Chris Gauker for discussion and advice about how to best express CA^A.

of modal properties then can be identified by the proper description of the type of event which counts as their manifestation.

Another important feature of the type of event the occurrence of which manifests an ability is that, although it has to involve *o*, it cannot always be described as '*o*'s φ-ing'. For many abilities' exercise does not entail their bearer's change. Surely, if an object *o* is fragile, *it* would suddenly fall apart if it were in certain circumstances. But there are many other abilities the exercise of which does not involve a change *in* the object to which we ascribe the ability. A hard object has the ability to break a fragile object, but nothing needs to happen *to that object* when it does so. It is only that the object manifests its hardness when it is involved in the event which results in a fragile object's sudden disintegration.

However, as we have seen in the previous chapter, there is another dimension of generality which can only be identified with reference to the conditions that need to be *actually* satisfied in order to correctly ascribe abilities – with the purpose to ground contingent possibilities – to an object. As we have seen, there is a difference between specificity as applied to the description of the manifestation event (which is analogous to the case of determinate/determinable relationship) and the specificity of abilities in this other dimension. If something is blue or less than 10 kilograms, it must also have a more specific shade (of blue) or a more determinate weight. If someone is able to speak a language, there must be at least one specific language one can speak. But, in another dimension of specificity, we can ascribe a less specific ability without assuming that the object has to have the corresponding more specific ability which would be manifested by the same type of event.

When we ascribe a maximally specific ability, we assume that all, known or unknown, conditions of its exercise are actually satisfied, except the occurrence of what is often called the 'stimulus' event. Thus, maximally specific abilities cannot be masked at all. 'Masking' an object's ability just means that its maximally specific ability is lost. However, what the examples of 'masking' and 'antidotes' indicate is that, on many occasions, we can ascribe some more generic ability to an object *without* assuming that it possesses the specific one.

Hence, we need some guideline that assists the selection of the specific conditions which must be *explicitly* included in the antecedent of the conditional that specifies an ability. Our guideline is this: the *less* specific the ability that we want to ascribe is, the *more* conditions need to be included in the antecedent of the conditional that specifies it. Examples of masks and antidotes show us which those other conditions might be. Hence, in a certain sense, an ability's degree of generality in this dimension is inversely related to how detailed the specification that we use in the antecedent of the conditional is. By explicitly including more and more conditions into the antecedent of the conditional, we can identify more and more generic abilities, which can ground more and more abstract possibilities.

Suppose I wonder about the possibilities that a mug, soon to be delivered to me by mail, breaks. I hope the mug does not have the specific ability to break during delivery because it is wrapped by some protective material. But I also think that it would break if it were transported in the way it is *and* it were not covered by the protective material. This conditional specifies (roughly) an ability which grounds a more abstract possibility in the sense that its exercise requires the satisfaction of more actually non-satisfied conditions than the exercise of an ability specified by a simpler conditional does. But it *does* identify a possibility about the mug's breaking. It is that possibility which makes it rational to actually wrap the mug with protective material.

This means that the more abstract possibilities we want to identify, the more conditions need to be explicitly specified in the set of conditions C in the specification of the corresponding ability. And inversely, the less abstract possibility we want to identify, the fewer conditions we need to explicitly identify in the conditional's antecedent. It is not going to be true of the mug in my previous example that it would break if it were dropped. But neither is it true of it that it has the *specific ability* to break in *those actual* circumstances. What is true of it, however, is that it would break if it were dropped and the protective material were removed. But then, it is also true that it *has the ability to break* identified with reference to *those* circumstances. In either way, the link between the ability at the relevant level of generality and the truth of the conditional with the appropriate set C in its antecedent is not severed.

Why finks and masks are different

As we have seen, 'reverse finks' raise a challenge for TCA that seems to be similar to masks and antidotes. When 'reverse finks' are present, an object can have a disposition or ability to φ at a certain time t; nonetheless, the corresponding counterfactual is false. The object would not φ even if it were in C at t.

Thus, it might seem that we can answer the problem raised by reverse finks in the same way we answered the problem related to masks and antidotes. On the one hand, the presence of a 'reverse fink' renders the conditional false; but whenever the conditional is false, the object lacks the corresponding *specific* ability too. On the other hand, if our aim is to identify a more generic ability in order to ground some more abstract possibility, then it is natural to include in the antecedent of the conditionals the supposition that 'reverse finks' are absent.[12]

However, this cannot be the correct answer when finks, rather than 'reverse-finks', are present. In the presence of masks and antidotes an object can *retain* the more general ability to φ even in the circumstances in which it lacks the more specific one. Thus, the conditional is true because the object does have an ability at the relevant level of generality. But in the case of finks, the counterfactual is true of an object even if the object *lacks* the relevant ability. This means that the corresponding conditional cannot be entailed by the object's having the ability. Thus, the possibility of finks raises a fundamentally different philosophical problem about the connection of ability-ascriptions and the truth of the corresponding conditionals than the phenomenon of masks and antidotes does.

There are many philosophers who agree with this: the possibility of finks and the possibility of masks raise different difficulties for the conditional analysis. But their reason to distinguish the two does not seem to me right . These philosophers, following David Lewis, consider modal properties as intrinsic properties of objects and hold that masks are related to the *extrinsic* conditions in which they are manifested.[13] In contrast, in their view, the problem of finks concerns the *intrinsic* nature of the object to which the disposition is ascribed. It is for this reason that the phenomenon of masking is different from the phenomenon of 'finking'. The former concerns the extrinsic conditions of manifestations, while the latter concerns the intrinsic properties of objects.[14]

As a consequence, David Lewis holds that we can solve the problem of masks by introducing explicitly dispositional locutions. As we have seen in Chapter 3, we cannot.[15] With the help of such explicit locutions we can distinguish dispositions from abilities and introduce some yet unspecified modal properties. But we can specify such properties only through their connections to counterfactual

---

[12] As for instance Choi does, in Choi (2008).
[13] In fact, Lewis has a somewhat odd argument for intrinsicness of dispositions (Lewis 1999/1997, 147). Imagine, says Lewis, that Willie has the extrinsic property of being protected by a big and strong brother. This does not mean that Willie himself is a dangerous man to mess with (where being dangerous is a disposition). But if we generalize this argument, we should conclude, by parity of reasoning, that Hitler was not a dangerous guy, since he was not particularly big and strong. Moreover, even if Willie is not dangerous (to mess with), Willie's brother must have the disposition to protect *him* (and not you or me), so Willie's brother's disposition must be extrinsic. If Willie dies, his brother's disposition is lost.
[14] Michael Fara, for instance, distinguishes the two kinds of problems in this way; see Fara (2005). There is a subsequent debate over whether masks or finks can be intrinsic. See, among others, Choi (2006) and Clarke (2008). However, if abilities can be extrinsic, then the debate about the possibility of intrinsic finks seems largely irrelevant.
[15] As was first argued by Bird (1998).

conditionals. Moreover, as I also argued there, most specific abilities are extrinsic. Thus, Lewis's distinction, even if it is applicable to the case of (assumedly always intrinsic) dispositions, does not apply in the context of abilities.

Despite this, Lewis does pinpoint a fundamental difference between masks and finks. Finks, as he notes, *change the very property* that we want to specify with the help of a counterfactual conditional during the process of its would-be exercise or manifestation. The phenomenon of masks and antidotes illustrates how sensitive abilities at different levels of generality might be to their bearers' circumstances. The phenomenon of finks and reverse finks, in contrast, occurs because the ascription of modal properties has a dynamic aspect: an object can change with respect to their modal properties in the very circumstances in which they would be exercised.

This change might be the consequence of another object's ability, as in Martin's original example about the 'electro-fink'. Finks are, more generally, objects with an ability $α^*$ to change another object's ability $α$ in the same circumstances in which $α$ would be exercised, if $α^*$ were absent. But, most importantly, the phenomenon of finking does not require the presence of objects with such abilities. Many specific abilities can change as a result of some change in the object's extrinsic circumstances. What makes finks possible is that an object can acquire and lose a modal property at any time, irrespective of the specific source of that change.[16]

'Finking an ability', in the philosophically interesting sense, means making an object acquire the ability just at the time and circumstances in which it is about to be exercised; 'reverse finking' means depriving an object of its ability at the time when it is about to be exercised. Finks and masks are related in the following sense. As we have seen, maximally specific abilities are most often extrinsic. Consequently, such abilities *cannot be masked*, but they can be very easily 'finked'. The more specific an ability is, the easier it is 'to fink' it. We do not need to install any special device in order to change an object's specific ability at the time when it is about to be exercised.

To illustrate, consider a hostage who is kept in a room which has only one firmly locked exit. This person does not have the specific ability to leave that room in those circumstances, even if she has a 'masked intrinsic ability' to leave it. When the room is unlocked, the hostage can regain its ability to leave. But if a vigilant guard suddenly locks the exit when and *only when* the hostage takes a

---

[16] More precisely, as I shall argue later in this chapter, we need to distinguish two kinds of change: some changes are the results of exercising abilities, whereas others are changes in having an ability.

few steps towards the door in order to leave the room – that is, when she begins to exercise her ability to leave it – she loses her specific ability, even if this change is extrinsic to her. As long as the door is unlocked, the hostage has the ability to leave it; but that ability is 'finkish' because some extrinsic changes at the time when it is about to be exercised can change the object's – or, in this case, person's – ability itself.

Objects' more generic, and hence less extrinsic, abilities can, of course, change as well in the moment when they are about to be exercised. It is logically possible that someone gets paralyzed only in the moment when she actually tries to walk; or that a paralyzed person acquires the ability to walk only in the moment when she decides that she will walk. The ability to walk when one tries is a very generic ability. But even this ability might be lost or gained in the circumstances in which it would be exercised.

All this poses an obvious difficulty for CA^A. The antecedent of the conditional that aims to specify an ability does so by identifying the conditions in which it would be exercised. It is possible that all those conditions are satisfied, let us say, at time t_0, but some of them change during the time between t_0 and t_1, the latter being the time by which the results of the exercise of the relevant ability are supposed to occur. Alternatively, the object might fail to have an ability, but can acquire it as a response to the mere appearance of the circumstances that are necessary for its exercise. Finking an object's ability means *changing* the object in respect of which abilities it has.

This crucial difference between finks and masks might be masked – excuse for the pun – by Charles Martin's original example, in which an extrinsic device is responsible for the intrinsic change in the object. But, metaphysically speaking, it is the change itself, and not its origin, what matters. Thus, we need to slightly revise CA^A in order to accommodate the possibility of 'finkish changes' of *abilities*. Again, dispositions are a somewhat different matter since, being a different type of modal properties, they are connected to different forms of conditionals. I shall discuss dispositions and conditionals in the last section of this chapter.

Losing and gaining abilities

Martin claims that the phenomenon of finks proves that counterfactual conditionals are 'only clumsy and inexact linguistic gestures to dispositions'. Somewhat later David Lewis claims that the failure of conditional analysis would prove the 'irreducibility of dispositionality'.[17] In fact, I doubt that these consequences follow. The failure of the analysis need not compel one to accept 'irreducible' first-order modal properties in one's ontology. And, more importantly for my purposes, granting that there are such properties does not mean that the ascription of abilities and the truth of certain conditionals are not logically connected.

Lewis aims to answer the problem of finks in reductivist spirit. He observes that the presence of finks and reverse finks can change an object's disposition after the 'stimulus condition' of the disposition occurs, but before the process of manifestation has been completed. But if the phenomenon of finks (and reverse finks) is essentially related to the possibility of such changes, then, in order to answer the difficulty raised by it, all we need to do is to add another supposition to the antecedent of the relevant conditional. On that supposition, a certain type of event Φ would occur, if *some property of the object* were retained (or remained unacquired) until its disposition becomes manifest; or until the exercise of its ability is completed.

But what is exactly that property? Taking his cue from earlier causal accounts of dispositional properties, Lewis assumes that every disposition must have a 'non-dispositional causal basis' – 'some intrinsic property B' – the presence of which, together with the stimulus event, would be causally sufficient for the occurrence of a disposition's manifestation.[18] For instance, when we ascribe 'fragility' to a solid object, then we must assume that it would break if it were struck *because*, given the crystalline structure in which the particles composing the object are arranged, the hitting would cause its breaking.

Assuming then that the manifestation of a modal property must involve a causal process that requires a 'causal basis', Lewis suggests amending the antecedent of the conditional with the supposition that, in the circumstances of manifestation, the object's intrinsic property B were retained until the occurrence of the manifestation event. This is Lewis's influential 'reformed conditional analysis' (or LRCA for short).

(LRCA) Something o has a disposition to r as a response to s at time t if and only if, for some intrinsic property B that o has at t and for some time t' after t, if o

---

[17] Martin (1994, 8) and Lewis (1999/1997, 139). This is, for instance, Molnar's view; see Molnar (2003, 82–98). This view is often identified with realism about powers. For important exceptions, see Mellor (1974, 2000) and Mumford (1998).
[18] About the idea that dispositions must have 'causal bases' see Armstrong (1973, 12–14), Mackie (1977), Prior at al. (1982).

were to undergo stimulus *s* at time *t* and retain property *B* until time *t'*, then *s* and *o*'s having *B* would jointly cause *o*'s giving response *r*.[19]

This analysis differs from the TCA in two crucial respects. First, it introduces an explicit reference to a 'causal basis', the necessary existence of which was, in some earlier accounts, only a metaphysical assumption, but not *part of the semantic analysis itself*. Lewis's analysis, in contrast, explicitly includes in the antecedent of the conditional the supposition that the causal basis is retained until the completion of the response-process. This means that a metaphysical assumption is already built into the semantics of modal properties. Admittedly, then, the semantics of modal properties cannot be independent of our ontological commitments.

Second, although this is hardly ever mentioned, Lewis reinterprets the very notion of manifestation. The analysis implies that what manifests a disposition is not the occurrence of a type of event, but *the causing of the event* by some intrinsic property of the object together with the occurrence of the 'stimulus condition'. I shall discuss the significance of this second, often neglected, feature of LRCA later in Chapter 6. Nonetheless, it is important to note right here that Lewis – as many other philosophers – understands dispositions as *causal* in the sense that their ascription presupposes an implicit claim about some potential causal processes. However, as I shall argue there, even if the *exercise of an ability* (or the manifestation of a disposition) does sometimes involve some such process, this does not mean that the analysis must aim to specify a modal property with reference to such processes rather than with reference to the events which are their results. What manifests fragility, one might insist, is the splintering of the object, not the causal process (if any) leading to the splintering.

Lewis mentions only *one* condition of manifestation, which he calls the 'stimulus' for manifestation, and he understands manifestation as a 'response' to a stimulus. This is, again, the consequence of understanding dispositions as potential causal processes that must have *a* cause. However, as we have seen, a generic ability cannot be specified with the help of a single condition (and this is, as we shall see later in the chapter, true of many dispositions as well). Only maximally specific abilities can be specified with the help of a single supposition.

Consider the generic ability of being ignitable. An object is ignitable if (very roughly, as a first approximation) it would catch fire if it were heated in the presence of oxygen. What is the 'stimulus condition' here? Well, it may depend

---

[19] I changed the notations and somewhat simplified Lewis's own formulation, but, in essence, I just recapitulate Lewis's proposal in Lewis (1999/1997, 149).

on which specific ability of the object we are interested in. If it is an already heated body, then it is the occurrence of the (appropriate amount of) oxygen in the object's circumstances. The object has the specific ability to catch fire in the sense that it would catch fire if oxygen were present in an amount m. If oxygen is already present, the same object has the specific ability to catch fire in the sense that it would catch fire if it were heated. We might distinguish these two specific abilities from each other. But objects that have either of them must have the same more generic ability which is usually ascribed to an object when we say that it is ignitable.[20]

Viewed from this perspective, Lewis's analysis is only a reinterpretation, and not a 'reform' of TCA. According to this interpretation, we need to include in the antecedent of the conditional aiming to specify a modal property the supposition that, in addition to the presence of the assumedly *extrinsic* 'stimulus condition(s)', some *intrinsic* property of the object is present as well. This is not a particularly surprising refinement of TCA. One important condition in which a cup of hot water has the ability to – or is disposed to – dissolve a sugar cube is that it is not already completely saturated with sugar; which seems to be its intrinsic property just as its temperature is.

Thus, some conditions that are intrinsic to an ability's bearer have always been tacitly assumed as necessary for the exercise of an ability. However, and much more importantly, if the aim of the analysis is to provide an account of 'dispositionality' without any reference to modal properties, then the relevant 'intrinsic basis' must be taken to be non-dispositional and hence *distinct* from the ability (or disposition) of which it is the basis.

It is, of course, controversial how properties which are usually mentioned as examples of 'causal bases' can be nonmodal. Macroscopic modal properties are often explained with reference to microscopic ones, but the relevant microscopic properties seem to be as modal as the macroscopic which they explain.[21] Thus, even if a disposition or an ability does have a 'causal basis', this will not help

---

[20] Lewis can of course respond that the work is done by his explicitly dispositional locutions. We may say that something is ignitable if it were ignited in the presence of oxygen as a response to being heated. But first, as I have already argued in Chapter 2, the object's ability to ignite is not the same as its disposition to ignite. Second, and more importantly in the present context, we could not solve the problem under discussion with this move; we can only shift it to the level of explicitly dispositional locutions. For why not to specify being ignitable as being disposed to ignite in sufficiently high temperature as a response to the occurrence of oxygen? Would it be a different disposition than what we normally mean by 'ignitable'? If not, it is much simpler to assume that there are many conditions from which we single out only one as the typical 'stimulus condition' on some pragmatic ground.

[21] The microphysical properties, typically mentioned as examples for 'causal basis', do seem to be modal as Mellor (1974), Ellis and Lierse (1994), Ellis (1999) and Molnar (2003) or McKitrick (2009), among many others, have shown.

'analysing away' modal properties. In fact, it seems that the 'more down we go' – that is to say, the smaller constituents of matter we investigate – the less likely it is that we can identify a property nonmodally.[22] Thus, the undeniable truth that the modal properties of macroscopic objects must have some microscopic 'basis' is just another metaphysical problem about the structural connection among certain modal properties. It will not help 'analysing away' modal properties in terms of nonmodal ones.[23]

Finally, Lewis assumes that the 'causal basis' of modal properties must be an intrinsic property of the object to which we ascribe the disposition. This might be reasonable, if we insist that abilities and dispositions themselves must be intrinsic. However, as our earlier considerations have shown, most abilities are not intrinsic, but variably extrinsic. And so are, though to a lesser extent, at least some dispositions. Hence their 'causal basis', if they have any, must be extrinsic too.[24]

This means that Lewis's analysis is based on some assumptions which restrain its applicability. Nonetheless, Lewis's basic insight that the problem of finks concerns the possibility of *change* of an object's ability at the beginning, or during the process, of its exercise, as opposed to the problem of masks and antidotes which concerns the conditions in which the ability can be correctly ascribed, is right and important. Consequently, we need to amend CA^A in order to answer the difficulty raised by the possibility of 'finkish change' of abilities.

A nonreductive conditional analysis of abilities

When Martin introduced the example of finks, his aim was to criticize 'eliminativist' or 'reductivist' analyses of powers and dispositions. But Martin has not shown why some version of the conditional analysis cannot be correct if our aim is not to reduce or eliminate modal properties. In fact, he says that 'there can be no conditional which is both logically equivalent to a categorical ascription *and* such as to support the elimination of power or dispositional

---

[22] The reason for this is the essential connection between these properties and the laws in which they figure. See Loewer (2007) and Huoranszki (2019).
[23] Moreover, the metaphysical nature of the connection between modal properties and their 'non-dispositional causal base' is hard to cash out. Lewis struggles with understanding the connection at one point but seems to admit partial defeat, Lewis (1997/1999, 143–4).
[24] About the first possibility, see McKitrick (2003); about the second, Smith (1977). Further, some finks may be intrinsic to their bearer. And if such finks can be removed only together with the intrinsic causal base, Lewis's analysis fails. About the possibility of intrinsic finks see again Clarke (2008, 2010), and Choi (2008).

predicates' (Martin 1994, 6, emphasis in the original). This might be right, but it does not imply, as Martin claims, that counterfactual conditionals are only 'clumsy gestures'.

The important point is that the 'analysis' of modal properties – that is to say, an account of the connection between their ascription and the truth of certain conditionals – need not be elimination or reduction. And if our analysis aims to specify an ability, then it is natural to add to the antecedent of the relevant conditional that the ability *will not be lost* until its exercise has been completed (Mellor 2000, 7–8). Of course, in some cases, the property is lost at the end of the process, often together with its bearer. No vase can remain fragile after it has been broken; and no sugar cube is soluble after it has been dissolved. But objects can retain their ability to break or to dissolve (without residue) until the process which constitutes their exercise is completed.[25]

Thus, as a response to the possibility of 'finking', we can modify CA^A and use the following schema:

> (RCA^A) For any object (or substance, or ensemble of objects) o and time t, there is a time $t' \geq t$ such that: o has the ability A at t if and only if an event of type Φ involving o would occur at t', if o were in conditions C at t and it did not change with respect to having or lacking A until (the process leading to) the occurrence of event type Φ is complete.

As before, to apply this schema, we first substitute letter 'Φ' with a description of a type of event and letter 'C' with a description of a set of conditions. And this choice is guided by which, and how generic, ability we want to specify when we substitute letter 'A' with a suitable term or description.

For instance, a sugar cube has the ability to dissolve in water at t_0, if and only if it would (fully) dissolve until t_1, if it were immersed in some not yet saturated, not very cold, and so on water *and* it retained that ability until it (fully) dissolves. My guess is that the satisfaction of the last condition has always been tacitly assumed by TCA. But we can also explicitly include it in the antecedent of the conditionals which we use to specify abilities.

This analysis does not assume that abilities *must* have some 'intrinsic causal base'. Thus, its scope is much wider than the scope of Lewis's account. Of course, there is nothing in the analysis that would make it incompatible with the assumption that *some* abilities do have an intrinsic base (whether the base must be 'causal' is a further issue, to which I shall return in the Chapter 6). As

---

[25] Not every ability's exercise involves a process, as we shall see in Chapter 6. However, abilities the exercise of which requires no process cannot be 'finked', so this is irrelevant in the present context.

a consequence, our proposed analysis is compatible with the extrinsicness of abilities. And this is important, for otherwise we could not ascribe abilities which ground contingent possibilities at different levels of abstraction.

One might object that this is not really an analysis. For the proposed amendment seems to render it 'circular', given that it includes a reference to the very property to be analysed in the *analysans*. However, this does not mean that the analysis must be 'circular' in the sense of being unilluminating. In fact, to stretch the spatial metaphor further, the analysis is entirely 'linear'.

The aim of our analysis is to specify an ability which an object (system of objects or substance, etc.) has at a moment. Many abilities are dynamic properties, the possession of which is essentially linked to what can or cannot happen in or with the object, or in its environment, in the future. The right-hand side of RCA^A only says that a certain sort of event would happen only if the object did not lose that property until the event happens. This seems to be an almost trivial condition, tacitly assumed in almost any account of dynamical properties.

Here is a test case to illustrate how innocent adding this further condition to the antecedent of the conditional is. Suppose I want to identify a specific ability, call it 'F', of a given macroscopic solid object s at time t_0 by saying that

s has F at t_0 if and only if it would suddenly disintegrate before t_1, if force f were exerted on its surface at region r from angle α at t_0 and if it did not change with respect to having or lacking F until t_1.

Do you have any difficulty with understanding which ability the letter 'F' represents? If not, then the analysis cannot be 'circular' in the sense that it is unable to identify a specific modal property. Thus, that 'F' appears also on the right-hand side of the biconditional does not make our analysis useless as *a semantic device* to specify abilities.

As a rough approximation, this application of the schema RCA^A says that s possesses F at t_0, if and only if s would φ in C while not losing/acquiring F after t_0 and before t_1. In order to specify F, it is enough that we understand what φ and C are. What is added is only a characterization of the connection between φ and C; a Φ-type of event would occur only if both C and F were present. But, again, this has been assumed already.

In fact, as I mentioned at the beginning of the chapter, one of the fundamental problems with early versions of the conditional analysis was that they did not attribute any role to *objects* and their *properties* in their account of the meaning of 'disposition terms'. They interpreted sentences containing disposition terms as a way to express a functional relation between occurrences understood as manifestation-conditions and occurrences understood as manifestations. The reason for this omission was that early analyses of dispositions presupposed the truth of phenomenalism which entails that *persisting objects as well as properties themselves* should be understood as functional relations among observation-events.

However, if we are ready to abandon phenomenalism, which I think we have many independent reasons to do, then the analysis of the content of an ability – or even a disposition – can include in the *analysans* an anaphoric reference to the property analysed. And the correct use of this referential device does not assume that we understand the specific content of the property which we want to analyse *before* the analysis. Knowing that *someone* is waiting for you at the airport and that *s/he* will take you to the hotel does not assume prior knowledge about who the person is.

It is interesting to note that, although in a less transparent way, the right-hand side of Lewis's analysis also contains an implicit reference to the modal property to be analysed. According to my suggestion, the antecedent of the conditional mentions the presence of both C and the dynamic property F. According to Lewis's suggested revision, the antecedent of the conditional includes the suppositions that both C (or, rather, a 'stimulus-event') and F's 'intrinsic causal basis' is present. But as far as the specification of a modal property is concerned, the latter formulation just introduces a further factor without helping to specify the relevant property.

What I mean is this. If reference to the retention of the ability to be analysed in the *analysans* is an objection to my proposed version of the conditional analysis, then it is also an objection to Lewis's analysis. For Lewis's analysis too contains an implicit reference to the modal property to be analysed in the *analysans*. In his analysis, one of the conditions of manifestation of an object's disposition is that it retains 'some intrinsic property *B*' until the manifestation has been completed. What is that property? Obviously, it must be the property that serves as the 'causal basis' of *the disposition to be analysed*.

Ordinary macroscopic objects have many abilities or dispositions at the same time. A knife made of stainless steel, for example, can have the ability to conduct electricity, to resist rusting, as well as the ability to cut bread (or the more generic ability to cut or scratch objects made of material less hard than steel). These are obviously not the same abilities, since certain (kind of) objects can have one of them without having the other. Soft objects – like my body – can be good conductors, and hard objects – like a piece of rock – can be very bad ones.

So how can we single out *the* 'intrinsic property B' the retention of which is necessary in order for the ability to become manifest? It seems that we must *stipulate* that, whatever property is the 'causal basis' of the disposition to be analysed, *that property* must be retained. Of course, if we assume that that property is distinct form the modal property itself, then there might be ways to identify it independently of the disposition or ability; even if this does not seem to be possible in every case.[26] But the relevant question is whether we can identify it *as the property that is the basis* of the relevant ability (or disposition) without some implicit reference to the ability (or disposition) itself.

Philosophers, who believe that there are not irreducibly modal properties and hence objects' dispositions or abilities must have some non-dispositional ground, often hold that dispositions are second-order properties of having some first-order nonmodal properties.[27] But this conviction is entirely independent of the phenomenon of finks or any other objections to the conditional analysis. And it is hard to see why an analysis which includes the condition that the ability A is retained until its exercise has been completed is less informative or 'more circular' than the one with the condition that *whatever property is the basis of the ability* A, it is retained until the occurrence of the proper 'response'. If there is a difference between the two, it is certainly the second, which introduces yet another unknown property that can be identified only through reference to the analysed one, which seems to be less informative.

'Achilles heels', accidents and the conditional analysis of dispositions

A hardened glass can fall on the ground in such an unfortunate manner that it breaks, even if it was not disposed to. The glass might have an 'Achilles heel'. It has been argued that the phenomenon of 'Achilles heels' raises an especially difficult problem for the conditional analysis of dispositions. We can accommodate the phenomenon of masks and antidotes into our analysis by explicitly introducing

---

[26] In many cases, we identify a microscopic property precisely by its potential contribution to the changes of observable objects and systems. No theory about microphysical particles could exist unless such particles had the ability to leave marks of their tracks in a cloud- or bubble-chamber. About the importance of this for an account of the contingency of physics, see Huoranszki (2019).

[27] Some philosophers hold that the 'reductive' relation between the two types of properties must be identity, but that is an incredible doctrine. As Lewis notes, free electrons are the causal basis for both metal's conductivity and their distinctive colours, but metals' conductivity and their colour are not the same properties. See Menzies (1988).

more conditions into the antecedent of the conditional which specifies an object's (or a person's, substance's, system's) disposition. It has been claimed, however, that this strategy does not apply in the case of 'Achilles heels'. It requires some explanation, why.

The first thing to note about 'Achilles heels' is that the kind of phenomenon that they describe is much less pervasive than the phenomenon of masks and antidotes is. As I shall argue in the next chapter, virtually every ability except the maximally specific ones can be masked (maximally specific abilities, in contrast, are very easy to fink). But objects (not to mention substances) have many abilities and dispositions which are immune to the 'Achilles-heel' problems.

Flexible or fragile objects can unexpectedly break or remain intact in circumstances in which we would normally expect them, respectively, to regain their original shape or to break. But it seems that they cannot unexpectedly dissolve or fail to dissolve in some acid; and they cannot unexpectedly conduct, or resist, heat or electricity. Elementary particles' dispositions may be masked, but how can they have 'Achilles heels'? Achilles heels are modelled on fragility, in both the literal and the metaphorical sense, but fragility, hardness and flexibility are a very special kind of dispositions, and what we learn about them might not be generalizable at all.

More importantly, as far as abilities are concerned, Achilles heels can be considered as peculiar cases of masking. The standard examples of masks that I have discussed so far are extrinsic factors which can thwart the exercise of an ability or the manifestation of a disposition in circumstances in which, if they were absent, the ability would be exercised. True, in the case of Achilles heels there is not any such *extrinsic* factor present. Yet, we can interpret Achilles heels as *intrinsic* masks. Since not every glass or every spring has an Achilles heel, there *must* be *some* intrinsic difference *in* the object which explains the difference between those objects that have Achilles heels and those which do not. And then, Achilles-heels work like intrinsic masks.

Here is an imaginary example to clarify what I mean. A glass's (as every solid object's) fragility is determined by the crystalline structure of the substance which constitutes it. Suppose we have a glass in which particles are homogeneously structured as S_f, except in one small region at its edge r^E where they are structured as S_h. Since, by our assumption, fragility is the result of the crystalline structure in which the constituent particles are arranged, this glass will have *two* relevant modal properties. (Of course, it will have many other modal properties unrelated to breaking, but this does not matter for our example.) It is *disposed* to break when it falls from a table on the floor; but it is *able* to retain its shape when it falls from the table on the floor. The two modal properties are compatible, because dispositions ground tendencies, and a tendency admits exceptions.

Now, as I argued in Chapter 2, the disposition to break need not imply that an object has the *specific* ability to break. The glass described earlier is not able to break in the specific circumstances in which it lands on its r^E. But it does have the ability to break when it falls in any other way. Even then, the glass has a more generic ability. It is such that it would break if it fell from a table and if it did not land on its region r^E. Landing r^E would mask the glass's generic ability (as well as its disposition) to break when it falls on solid ground from a certain distance. However, at the same time, the glass also has the more specific ability that it would break if it fell from a table and did not land on r^E.

Can we also say that landing on any other region than r^E masks the glass's ability to retain its integrity when it falls from the table? Yes, I think we can. As far as objects' *abilities* are concerned, intrinsic masks can be symmetric. For the conditions in which the ability to φ would be exercised can include reference to the absence of conditions in which the ability to resist φ-ing would be exercised. And inversely, the conditions in which the ability to resist φ-ing would be exercised can include reference to the absence of those conditions in which the ability to φ would be exercised.

However, dispositions are a different matter. If r^E is very small, it would be false to say that the glass is disposed to retain its integrity when it falls from the table. Dispositions entail tendencies and (assumedly fragile) glasses do not have a tendency to *resist* breaking in the circumstances in which the glass in our example did (at least if their constituent particles are structured in a like manner). It is not a response to this, that such glasses have the tendency to resist breaking when they land on r^E, for *there is not such tendency*: the very concept of Achilles heels rests on the assumption that such cases are exceptional, rare or accidental. This means that an object is *able* to behave in ways in which it is *not disposed* to.

Achilles-heels scenarios then describe exceptions: cases that deviate from the 'normal'. But again, this does not distinguish them from some masks. A nuclear power station's disposition to overheat is masked by a safety device, and, in this case, masking is the 'normal' state of affairs. In contrast, a compass's pointer's 'normal' disposition to turn towards the North Pole can be masked by a nearby magnet. The only difference between Achilles heels and this sort of accidental masks is that Achilles heels are *exceptions by definition*.

More interestingly, however, the phenomenon of Achilles heels and masks is closely related to the issue of dispositions and laws, which I discussed in

Chapter 2. There I argued that laws explain their instances only to the extent that the instantiation of a feature F or the occurrence of a type of event Φ is a manifestation of an object's or a system's disposition to display F or to φ in C. The possibility of Achilles heels shows that an object can be disposed to φ in circumstances C even if in *some* (actual or counterfactual) cases it fails to φ in C; or that an object can be disposed to display F even if, on some occasions, it fails to do so. Nomic generalizations' modality is compatible with *admissible exceptions* because objects having a disposition can have Achilles heels or accidental masks.

What makes exceptions admissible? What makes them admissible is that they count as *accidents* in a somewhat technical sense of the word. Tigers are genetically disposed to have four legs. Thus, whenever they have less – or more? – this must count as *an accident*. The term 'accident' ordinarily refers to something that is unexpected or 'deviant' in some way. But the concept of an accident can also be used in a more technical, metaphysical sense.[28] In that sense, accidents explain how the truth of a nomic generalization is compatible with some exceptions. Accidents, by their very nature, cannot be part of the specification of the circumstances in which a generalization holds; but that accidents are possible explains how *ceteris paribus* generalizations can be laws.

But further, and much more importantly, accidents can also help to identify the modality of nomic generalizations. As I argued in Chapter 2, it follows from the truth of a nomic generalization that *if o* exemplifies a natural kind K and hence it is disposed to φ in C, *then* it is not an accident that φ occurs. It is not an accident in the sense that the non-occurrence of φ in circumstances C *would be an accident*. That an event happens in virtue of a law *does not* entail the impossibility of its failing to occur. But whenever the occurrence of a Φ-type events in C is a nomological consequence of an object's instantiating K, *and* nonetheless the kind of event does not occur even if C does, then the object must either have an Achilles heel or some yet unidentified mask must have been present.

Accidents in this sense are relative to the instantiation of some natural kind. It is an accident when a tiger has only two legs. It is not an accident when a chicken does. It is an accident if a mug made of clay falls from a table, but it does not break. It is not an accident if a plastic mug falls in the same circumstances, and it does not break.

---

[28] Importantly though, I am not using 'accidents' in the sense as it was used by the scholastics either. In that sense, accidents are such *properties* that cannot be used for individuating particulars (e.g. hair colour and bodily posture are 'accidents' in this sense).

Accidents are not unexplainable. Just to the contrary, they positively *demand* explanation.[29] If someone asks why a certain animal has four legs, as opposed to six or eight, it is perfectly satisfactory to answer that it has four legs because it is a tiger (rather than being a fly or a spider). This does not seem to require any *further* explanation. Similarly, if it is asked why a person died, it is a perfectly satisfactory response that he died because she has ingested a certain amount of arsenic. However, we do need to explain why an animal has only three legs *despite* being a tiger. (The animal has lost one of his legs in a fight; or it had this or that gene atypical to tigers, etc.) Similarly, we need to explain why someone who has ingested arsenic did *not* die afterwards. The point is not that we always have a satisfactory explanation. The point is that it is always natural to require one.

Of course, we can also ask why tigers are quadruped (in the sense of being disposed to have four legs) or arsenic is a toxin (in the sense of being disposed to poison). But that is a question about why some *kind* of thing or substance has a certain property. Accidents are, by their nature, singular occurrences. And it does *not* require an explanation why they do *not* occur. Think of the Achilles heel of a glass. If the glass instantiates the nomic property of fragility, it requires no further explanation why it broke when it was dropped. What we may enquire what makes, in general, things fragile. But if an object *is* fragile and it did *not* break despite that it was dropped, this naturally calls for some explanation.[30]

Accidents then can be understood as a special type of contingent possibilities, the best examples of which might be Achilles heels, because they are exceptions to certain nomic tendencies by definition. In order for an accident to happen, the object must be *able* to display a certain form of behaviour (or to instantiate a certain feature) which it is *not* disposed to. A *ceteris paribus* generalization can express a law because it would be an accident if an object instantiating the relevant nomic property failed to display a feature or behave in a certain manner (in circumstances C).

This observation has important consequences regarding how the ascription of dispositions is connected to counterfactual conditionals. In Chapter 2, I argued that nomic dispositions are always kind dependent properties. The same

---

[29] For more on the connection between *ceteris paribus* laws and the condition of explicability of 'exceptions', see Pietrosky and Rey (1995).

[30] We are all familiar with the situation when children keep asking why-questions even when we believe everything has been said that needs to be said. What we want to teach them at that point is not that they should stop raising questions because it is impolite or indiscrete, and so on. What they must learn is rather a fact: the fact that certain questions need not be asked because, objectively speaking, there is no more explanatory demand. Or, alternatively, that certain questions need to be asked exactly because there is a further explanatory demand.

thing can have a disposition to φ as belonging to a kind, and, yet, fail the have a disposition to φ as belonging the another, more specific kind. To recall our earlier example, someone considered as belonging to the biological kind 'human' must have the more generic disposition to digest cow milk, even if (a) because of her special condition, she is not able to; and even if (b) considered as instantiating the medical kind 'lactose intolerant', she does not have the more specific disposition to digest cow milk. The sortal- or kind-relativity of dispositions must be reflected in our account of disposition ascriptions and conditionals.

As we have seen in the same chapter, we also need to distinguish non-teleological and teleological dispositions. These different sorts of dispositions are connected to two different conditionals. First then, I suggest the following general schema for conditionals to specify nomic non-teleological dispositions.

> (RCA^{DN}): For any object (or substance, or ensemble of objects) o and time t, there is a time $t' \geq t$ such that: o has the nomic disposition D at t if and only if an o-involving event of type Φ would occur more often than not at t', if o were or became a member of K or an instance of N and it were in circumstances type C at t.

In order to apply this schema for specific cases, we first need to substitute letter 'D' with some term or description of a disposition on the left-hand side of the biconditional, while on the right-hand side we need to substitute letters 'Φ' and 'C', respectively, with a description of some type of event and some type of conditions; and 'K' or 'N' with some term referring to a natural kind or a nomic (physical) property.

The RCA of nomic dispositions differs from the RCA of abilities in two important respects. First, as I argued in the previous chapter, nomic dispositions are relative to some kind (like being a tiger) or to their bearers' being instances of a certain physical property (having a certain mass or electric charge). Of course, when objects belong to a kind essentially, then the condition concerning kind membership is trivially satisfied. But, first, whether or not kind membership is essential to the identity of an object or even an organism is a contentious issue. As we will see later in the next chapter, it is certainly not essential, for instance, to the identity of particles. And further, and more importantly, many kinds are certainly not essential to their members. Lactose intolerance is a medical kind, but it is not essential to any human person to be or become lactose intolerant.

Second, and relatedly, dispositions ground behavioural tendencies, and not the possibilities of singular events. This is expressed by the counterfactual that certain type of events would happen more often than not to an object, if they belonged to a natural kind or instantiated a physical property and were in some specified circumstances.

For the sake of simplicity, I formulated the conditional in terms of objects (or organisms) and events. But it is supposed to apply to features of objects (substances, systems of objects) that they display during a period of time as well. And further, it is important to note that we can ascribe dispositions not only to individual members of a kind or instances of physical properties but also to such ensembles which constitute a species. Thus, to use our earlier example again, tigers are disposed to have for legs if and only if more often than not they would have four legs if they had been born with a certain genetic structure. If having certain features is a *disposition* of an object, then there must be some conditions in which objects belonging to a certain kind must display that feature. I shall return to the significance of this claim in the next chapter.

As far as teleological dispositions are concerned, we ascribe them when a kind of behaviour is performed because it tends to satisfy some objects' or agents' goals. The following schema can be used to explain the link between the ascription of teleological dispositions and conditionals.

> (RCA^{DT}): For any object (agent or person) o and time t, there is a time $t' \geq t$ such that: o has the teleological disposition D at t if and only if o would display a Φ-type behaviour at t' more often than not in order to satisfy Θ, if o were in conditions C at t.

As before, to apply this schema for specific cases, we first need to substitute letter 'D' with some term or description of a teleological disposition on the left-hand side of the biconditional, while on the right-hand side we need to substitute letters 'Φ' and 'C', respectively, with a description of a type of event and a type of conditions; and 'Θ' with a description of some natural or acquired goal.

The relevant Φ-type behaviour can be some simple movement, but can be as complex as migrating, hunting or building a nest sometimes is. Some teleological dispositions are nomic in the sense that particulars (most typically biological organisms) have the goal Θ in virtue of being a member of a natural kind. But this is not required in every case since different teleological dispositions can be learned or acquired among members of the same kind.

Importantly, however, we can ascribe a teleological disposition only if Θ is not a *very specific* aim, which an agent can have only on a particular occasion. Jill might go to a nearby cinema in order to satisfy her aim to see the most recent Oscar-winning production. This action, which is performed with the intention to satisfy a specific goal, is a manifestation of one of Jill's (actional) abilities, but it is not a manifestation of a disposition.

Jane can, however, acquire the intention to see *every* Oscar-winning movie. Thereby she may acquire a disposition, because that intention entails that she would go to a nearby cinema more often than not in order to see a new Oscar-winning film, if she learned about the new winner and if it were on the program, and so on. Thus, the ascription of a teleological disposition presupposes the ascription of some sort of *generic* aim (or, in the case of intentional action, 'policy'), which an animal or human agent can regularly try to satisfy by its behaviour.[31]

The conditional analysis of dispositions is not exposed to the problems generated by the phenomenon of Achilles heels, masks or antidotes. Achilles heels, masks or antidotes count either as exceptions or as accidents, and then they are compatible with the frequent enough occurrences of the event-type Φ (or of feature F) which manifest a disposition. If they occur too often or regularly, then an object just ceases to be disposed to φ. Even in the latter case, however, the object might have the more or less generic *ability* to φ; but this is, as we have seen, a different matter.

The same is true about the phenomenon of finks. If an object's disposition to φ were 'finked' regularly enough, then it would be a mistake to ascribe the *disposition* to φ to the object. For if an object would lose its disposition to φ too often during the process of manifestation, then – contrary to what Martin assumes – it would not be disposed to φ; and if it acquired a disposition often enough, then it would indeed be disposed to φ.

In Martin's example, the artificial *systems* which are ensembles of copper wires and electro-finks are *not* disposed to conduct electricity when the wire in them is live; and they are disposed to conduct electricity when the wire in them is dead. But copper wires as instantiating a natural kind are of course disposed to conduct electricity when they are live because they are not frequently (not to mention naturally) part of such systems.

Let me finish this chapter with a few remarks about my use of the notion of tendency in my account of dispositions. As noted earlier, I do not want to deny that the *word* 'tendency' is sometimes used in everyday language even in contexts in which the frequency of certain type of events is relatively low. What I do want to say is that, in the context when our aim is to identify objects' dispositions and

---

[31] Artificial or technological kinds are related to teleological dispositions too. Roughly, an object has such a disposition if and only if it would φ more often than not in order to achieve Θ if it belonged to a kind of artefacts KA and were in circumstances C.

to distinguish them from 'mere' abilities, that is, in those contexts in which we aim to distinguish that which can possibly happen from that which happens in virtue of a nomic regularity or of some habits, it is theoretically advantageous to rely on the 'more often than not' criterion.

However, it is important to note further that there are certain contexts in which it would be misleading to talk about tendencies and dispositions, although what we really mean is only causal relevance. For instance, one might want to claim that tobacco smoking tends to cause lung cancer even if it does not do it more often than not.[32] This would mean that tobacco has the following disposition: if smoked regularly, it would cause lung cancer. But I do not think that tobacco has any such disposition or that tobacco smoking has the *tendency* to cause lung cancer.

What is true is only that smoking tobacco *can* cause lung cancer; or more precisely: we know that it can contribute to the development of lung cancer. Since it does this in a significantly high percentage of cases, this is enough reason to avoid smoking it (at least if we want to avoid developing lung cancer). But this does not mean that we should ascribe such a disposition to tobacco. Not fastening a seatbelt can causally contribute to serious injury in case of many accidents. So, it *can* causally contribute to injury. That's good enough reason to fasten our belts. But it would sound very strange to claim that seatbelts have the disposition to cause serious accidents when not fastened; or that not-fastening seatbelts has a *tendency* to cause injury.

Finally, I would like to add that the criterion of frequency that I propose as an interpretation of tendency is very minimal indeed. In scientific practice, much stronger evidence is needed to ascribe a disposition to an object. For instance, normally we need more than 50 per cent of positive cases in order to be justified in believing that a medication is disposed to cure a certain illness. However, the crucial point is the following: it is one thing to show that some (kind of) things in some conditions *can* do something, and it is quite another to ascribe to them the disposition to do it.

---

[32] I thank my reviewer for bringing this case to my attention.

5

Modal properties and conditionals

The scope of the connection

In the previous chapter, I argued that we can specify abilities at various levels of generality through their logical connection to counterfactual conditionals. In this one, I shall continue to argue for the plausibility of my revised version of the conditional analysis.

First, I shall investigate further the issue concerning what we could, and should, expect from a philosophical analysis in general. And then, I shall try to demonstrate the philosophical significance of the conditional analysis of abilities by presenting brief examples about how it can provide a link between the ascriptions of modal properties on the one hand and our theoretical and practical reasoning on the other.

Further, I am going to show how easily we can explain some well-known logical peculiarities of the use of subjunctive conditionals if we understand them as logical consequences of abilities ascribed to objects, persons or substances. Then I shall argue that relying on my proposed version of the conditional analysis we can give an adequate account of the nature of non-deterministic processes, again, by linking object's abilities to the sort of conditional which, according to my account, their ascription entails.

Finally, I shall address the question of whether there could be any kind of modal properties the ascription of which is not linked to the truth of some conditional. I shall argue that it is a mistake to ascribe 'continuously exercised powers' to objects and that the real nature of 'spontaneously exercised abilities' can be specified only with the help of some conditional.

A modal-property analysis of counterfactuals

To further justify the use of conditionals in the analysis of modal properties, let me return to the question concerning the very idea of an analysis as it is used in philosophy. As already mentioned in the previous chapters, historically, the idea of analysis has often been associated with 'reduction'. The notion of 'reduction' was originally introduced into philosophy in the early twentieth century as a part of the verificationist program in semantics. The so-called reduction-sentences were introduced with the intent to connect statements containing 'theoretical terms', which include those that contain 'disposition terms', to statements which record observations, thereby making the former 'verifiable' and hence meaningful.[1]

In contemporary philosophy, however, reduction is considered as a part of an ontological project. Though popular, it is not easy to comprehend what exactly the purpose of that project is.[2] However, even if we could clearly understand the meaning of ontological reduction, philosophical analysis need not be reductive in either sense. One can give an analysis of the concept of a cause without 'reducing' it to non-causes. And one can offer an analysis about modal properties like abilities and dispositions, without trying to reduce them to the nonmodal.

What does then analysis do? Consider a standard example, the analysis of the concept 'bachelor'. Suppose that being a bachelor is 'analysed' as being an unmarried man. In this 'analysis' the relation between the property of being a bachelor and the property of being an unmarried man is not symmetric. For the fact that someone is unmarried or that someone is a man is not explained by, or grounded in, the fact that the person is a bachelor.[3] Rather, it is being unmarried and being male, which make it the case, or grounds, that someone is a bachelor.

Non-realists about modal properties interpret the conditional analysis of abilities and dispositions in an analogous way. Just like the truths that someone is a man and he is unmarried explain and ground the fact that someone is a bachelor, counterfactual conditionals explain or ground the ascription of modal properties. It is partly for this reason that the truth of counterfactual conditionals itself must be accounted for *without* any reference to modal properties (rather, most typically, in terms of possible worlds).

But those who do not deny that there are first-order modal properties can understand the relation between abilities and conditionals differently. Perhaps it is the possession of abilities as first-order modal properties which grounds contingent possibilities *through* their logical connection to counterfactual

---

[1] See Carnap (1936–37).
[2] As Michael Jubien acutely observes, 'if the concept under analysis has a certain characteristic feature, [. . .] then one would think that feature must also somehow be present in the analysans, or else the analysis could not be correct. From this perspective [seeking 'reductive' analysis] looks like the pursuit of magic.' See Jubien (2009, 95).
[3] These considerations have been inspired by Kit Fine's seminal paper on modality and essence, Fine (1994).

conditionals. This does not imply the rejection of a logical or conceptual link between the ascription of abilities on the one hand, and the truth of certain conditionals on the other. But it changes the order of dependence.

The conditional analysis of modal properties, as any other analysis, is expressed by the formula 'p if and only if q'. If 'if and only if' is interpreted as a logical connective, then it should be interpreted as a biconditional. Biconditionals are symmetric; if 'p if and only if q' is true, then 'q if and only if p' must also be true. But 'q defines p' or 'q is an analysis of p' are *not* symmetric relations. Being an unmarried man might be a definition of being a bachelor, but being a bachelor is not a definition or an analysis of being an unmarried man. Thus, the question is which *non-formal* relation the use of some biconditional in a definition or analysis expresses.

This question is not susceptible to a uniform answer. How we answer it depends on the specific concept that we want to define or analyse and the specific type of analysis that we are interested in. As a first step, we may say that the biconditional provides an analysis if the truth of the right-hand side proposition in the biconditional is necessary and sufficient for the truth of the left-hand side proposition. However, even if this is right, this is not illuminating since 'q is necessary and sufficient for p' sounds just like saying that 'q if and only if p'. Hence, the question about how to understand the asymmetry of analysis is not thereby explained because 'necessary and sufficient' is just as symmetric as 'if and only if' is.

In order to understand the relevant asymmetry, we should begin with the trivial logical observation that the truth of a biconditional (as the word suggests) is entailed by the truth of two conditionals. The proposition that 'p if and only if q' is logically equivalent with the proposition 'p if q, and q if p'. But the two conjuncts taken separately are not symmetric. And this seems to provide a clue to the explanation of the asymmetry in the analysis. For the analysis to be correct, both conjuncts must be true. But they must be true for very different reasons.

Let us see what this means in the context of the conditional analysis of abilities. Suppose we want to apply our schema for some object s which has the ability F between time t_0 and time t_1. If RCA^A is right, then the following two conditionals must be true as well:

(CC) **If** s has the ability F between t_0 and t_1 **then** event φ involving s would occur at t_1, if s were in conditions $c_1, c_2, \ldots, c_n$ at t_0.

And

(PC) If the counterfactual conditional [event φ involving s would occur at t_1, if s were in conditions $c_1, c_2, ..., c_n$ at t_0] is true, **then** s has the ability F between t_0 and t_1.

These conditionals follow from an application of RCA^A to a specific case. It is important to note though that they are true for different reasons. The second conditional entails that if the counterfactual is true, then we are entitled to *ascribe* the ability F to s during a period of time, and it also indicates the degree of generality of F. In contrast, the first conditional tells us under what conditions φ, that is to say a possible event, would occur. And hence it shows how the possession of an ability entails the existence of certain types of possibilities. It explains that the fact or state of affairs that s has the ability F ontologically grounds the (actual or possible) occurrence of a φ.

Compare this to the case of bachelors. It is the fact that a man is unmarried which grounds the truth that he is a bachelor. Indeed, it would be very strange if the fact that a man is unmarried were grounded in the fact that he is a bachelor. Whatever is the ground of someone's gender or his lack of a spouse, it is certainly not that he is a bachelor. In contrast, a counterfactual conditional is true *because* some object (or substance, ensemble of objects) o has a more or less specific ability. For nothing can have an ability or any other property *actually* unless some *counterfactual* is true.

The second conditional can help us in specifying the ability F by linking it to a description of an event and to the specific set of conditions in which that event would occur. Specifying an ability or disposition is not the same as ontologically grounding it; it is a semantic and cognitive task. Ontologically, it is an object's having an ability which explains why it can display a certain kind of behaviour, or elicit some type of changes in other objects, in certain circumstances.

Given all this, there is sense in which it is more adequate to talk about the modal-property analysis (MPA) of (some) counterfactual conditionals, than about the conditional analysis of modal properties. This can be indicated by changing the order of the propositions on the two sides of the biconditional. We might represent the MPA of counterfactual truth with the following conditional.

(MPA) The counterfactual conditional [φ would occur at t_1, if s were in conditions C at t_0] is true if and only if s has the modal property/ability F between t_0 and t_1.

This analysis expresses more adequately the ontological relation between abilities and possibilities than the traditional conditional analysis does, since the sequence of propositions on the two sides of the biconditional clearly indicates that the more or less abstract possibility of the occurrence of φ is ultimately grounded in the abilities – that is to say, in one sort of modal properties – that object *s* has. But since the original purpose of the analysis was to specify an ability at a certain level of generality rather than to explain the ontological relation between abilities and possibilities, overall, I consider RCA^A preferable to MPA. Moreover, RCA^A is compatible with an object's having an ability only for an instant, while MPA does not seem to be.

However, it is important to emphasize that we need the left to the right (CC) and the right to the left (PC) connection for different reasons. The former is grounded in an ontological relation. It expresses the truth that the possibility as specified by the counterfactual conditional is grounded in an object's (substance's, ensemble of objects', etc.) modal property. The latter is needed in order to introduce an ability into our language and thought by specifying it with the help of a conditional. Possibilities exist because objects have modal properties. But we can understand and think about specific modal properties and their relations to each other only by way of connecting them to certain counterfactual conditionals.

Some remarks on the logic of counterfactual conditionals

Counterfactual conditionals are not used only in the context of reasoning that involves or presupposes the ascription of modal properties. Nonetheless, the phenomenon of masking and the operation of antidotes offer a plausible explanation about some intuitive differences in the logic of material and subjunctive conditionals. Material conditionals are transitive, can be contraposed and they retain their truth if their antecedent is strengthened by a true proposition. It is generally – though not uniformly – accepted that these rules are not valid in those contexts in which we reason with subjunctive, or counterfactual, conditionals.

A material conditional will not change its truth-value if we add a further true proposition to its antecedent(s). It means that its antecedent can be 'strengthened'. But it is not true that if we add a further supposition to the antecedent of a counterfactual conditional, this never changes its truth-value. In the context of counterfactual reasoning antecedent strengthening is not valid, and hence transitivity is provably invalid.

It is also easy to see why counterfactual conditionals do not always contrapose. I want to show that it is the phenomenon of masks, antidotes and the variable degree of specificity of modal properties which are responsible for these distinctive features of the logic of subjunctive conditionals when they are used in reasoning which involves or presupposes the ascription of abilities.

Consider a vase's very specific ability to break in the actual circumstances in which it is *not* filled with any protective material. Suppose it is true of that vase that it would break if it were in circumstances which differ from the actual *only* to the extent that the vase falls from a table on which it actually stands. But it is not true that the vase has the ability to break in such circumstances which differ from the actual in *two* respects: that the vase falls *and* that it is filled with some protective material.

Similarly, if someone ingested (a certain amount of) arsenic at t, she would get poisoned afterwards. But if she ingested arsenic and then took antidotes, she would not. Thus, the phenomenon of masks and antidotes provides a simple explanation of why antecedent strengthening is not valid in the context of counterfactual reasoning. Antecedent strengthening is not valid because we can specify more generic abilities, which are entailed by the ascription of the more specific ones, only with the help of conditionals which include some reference to the *absence* of counteracting factors.

This means that in those cases in which the new supposition states the presence of some mask or antidote, 'strengthening' the original conditional would result in a conceptual contradiction. For masks and antidotes are, *by definition*, factors or conditions that thwart the exercise of some other abilities which would be exercised, only if they were absent. Thus, it is logically impossible that when a supposition about the presence of a mask or antidote is added to a true counterfactual conditional entailed by some ability, the new conditional be true.

If a counterfactual conditional is true then because it is the consequence of the ascription of an ability, antecedent strengthening cannot be *generally* valid. Whenever the antecedent of a true counterfactual conditional, which is entailed by the ascription of a specific ability, is 'strengthened' by a condition stating the presence of a mask or an antidote, the resulting counterfactual will be false. That antecedent strengthening is not a valid principle is not an unexplainable logical fact. It follows from the possibility of masks or antidotes and the logical connection between the ascription of abilities and the truth of certain conditionals.

The contraposition of the conditional 'If the vase fell from the table, it would break' is 'The vase would not break, if it did not fall from the table'. The latter conditional is obviously false, since the vase would break if, for instance, it did not fall from the table, but were struck strongly by a hammer. Thus, counterfactual conditionals do not contrapose, because the conditions of exercising an ability or manifesting a disposition can be characterized by more or less specifically, and the absence of a specific type of condition (falling from the table) does not exclude the presence of another specific type of condition (being hit by a hammer).

What about very generic type of manifestation conditions like 'exercising force f from a direction d on the surface s of the object o'? 'The vase would not break at t, if no force had been exerted on it' does seem to be true. After all, this just means that a vase, unlike a house of card, is a stable macroscopic object. Thus, *some* counterfactual conditionals do contrapose.

When C in the antecedent of the conditional specifies conditions that are both necessary and sufficient for the occurrence of φ, then it is true that if φ did not occur, then C would not occur either. Later in this chapter I shall explain why some very generic abilities which ground counterfactuals do contrapose. But counterfactuals entailed by more specific abilities do not, and this is enough to explain why contraposition is not a valid principle in the logic of counterfactual conditionals.

Finally, transitivity entails antecedent strengthening, so if antecedent strengthening is not valid, counterfactual conditionals cannot be transitive. If someone ingested arsenic and took antidotes, then she would ingest arsenic. If she ingested arsenic, then she would be poisoned. But it does not follow that if someone ingested arsenic and took antidotes, then she would be poisoned.[4]

That counterfactual conditionals are not transitive, do not contrapose and antecedent strengthening is not valid in their context is relatively uncontroversial. However, there is a further logical principle, which is obviously not valid in the context of material conditionals, but it *is* controversial whether it is valid in the context of counterfactual conditionals. And our commitment about the validity of this principle plays an especially important role in accounts of nondeterministic abilities, which I shall discuss in the next section.

Some conditionals can oppose to each other. They have the same antecedent, but logically incompatible consequents, like 'If p then q' and 'If p then not-q' do. If a material conditional's antecedent is false, then the conditionals with the opposite consequences can both be true. But counterfactual reasoning is reasoning with the supposition that the antecedent of the conditional is true, and then it seems that opposite counterfactual conditionals cannot be both true. Nonetheless, I shall argue in the next section that objects can, at the same time, have abilities which entail the truth of opposite counterfactual conditionals.

---

[4] About this see Bennett (2003, 160–1).

So far, I have argued that, by using counterfactual conditionals, we specify objects' abilities that ground *possibilities* at different levels of abstraction. However, many philosophers, following David Lewis – or perhaps David Hume – interpret 'would-conditionals' as expressing 'necessary connections', rather than possibilities.[5] So understood, if the counterfactual 'If C were the case, then φ would occur' is true, then C would in some sense *necessitate* φ; even if the sense of necessitation is 'less strict' than logical necessity is.

This interpretation of 'would-conditionals' is clearly reflected in Lewis's semantics for counterfactual conditionals. In Lewis's account, the antecedent of the conditional C necessitates the truth of its consequence φ in the sense that the consequent is *true in all* closest possible worlds in which C is true. Briefly, 'If C were the case, then φ would occur' is true if all C-worlds are also φ-world. In this system, possibilities are expressed by 'might-conditionals' like 'If C were the case, then ψ might occur'. A 'might-conditional' is true if the consequent of the conditional is *true in at least one* world in which its antecedent is true.

The introduction of 'might-conditionals' seems to account for some form of counterfactual indeterminacy. If Bizet and Verdi had been compatriots, it is not true that Bizet would have been Italian; neither is it true that he would not have been Italian. Thus, the conditional excluded middle seems to be invalid. But, according to Lewis, we can avoid this problem. For it is true that Bizet *might* have been Italian; which is compatible with the truth that he might not have been Italian. In some of the closest 'Bizet-Verdi-compatriots-worlds' Bizet is Italian; in some others, Verdi is French.

As some other philosophers, I am sceptical about the idea of 'might-conditionals'. First, it is interesting to note that the distinction between might-conditionals and would-conditionals is a peculiarity of English language. It is certainly not universal. In many other languages the Bizet-Verdi case could be expressed by modifying the would-counterfactual, and not by the use of an 'opposite auxiliary', which is supposed to be the 'modal pair' of 'would'. What is natural to say in such languages is that *perhaps* if Bizet and Verdi were compatriots, both would be Italian; *perhaps*, both would be French. This clearly expresses epistemic uncertainty, not metaphysical possibility. If all that we know (or suppose) is that two persons are compatriots, this does not provide sufficient information about their would-be joint nationality.[6]

---

[5] See Lewis (1973). For an excellent summary of the arguments for, and debates over, 'might-counterfactuals' see Bennett (2003), Chapter 11.
[6] It seems that in the Verdi-Bizet case, it is usually assumed that the context narrows the number of options down to two (French or Italian). But this is not necessary. Perhaps, if Verdi and Bizet were

Further, might-clauses can be detached from their conditional antecedent in a way in which would-clauses cannot be. 'Verdi might have been French' *is* indeed a way to express a possibility. But 'Verdi might have been French' is not the *consequent* of any interesting counterfactual conditional. And if we interpret this sentence as expressing a metaphysical rather than an epistemic possibility, then it *does* entail the truth of some would-conditionals. It entails that Verdi would have been French, if he had changed (or decided to change) his nationality from Italian to French. This is certainly an ability Verdi had. But might Verdi have been born French? Verdi himself did not have this ability. That possibility depends on whether his parents could have changed their nationality, or at least moved to France, before Verdi was born. If they had those abilities, Verdi might have been born as French. *Mutatis mutandis*, the same is true about Bizet's nationality.

Nonetheless, when we are interested in such possibilities, it is more natural to say that Verdi *could* have been French, rather than saying that he *might* have been one. That 'might' expresses epistemic uncertainty rather than metaphysical possibility is also confirmed by the fact that its natural 'opposite' in English is 'must'.[7] When I say 'It must be the case', I express something like 'I am certain that it is the case'; whereas by saying 'It might be the case', what I express is 'I am not certain that it is not the case'.

Abilities and nondeterministic processes

The Verdi-Bizet case can be interpreted both as epistemic uncertainty and as metaphysical possibility. However, there are cases which seem to clearly involve metaphysical possibilities and which are not related to the issue of excluded middle. If we assume that throwing a fair coin is a genuinely indeterministic process, then the coin must have the ability to land heads up and the ability to land tails up. But, on the account I am arguing for, these abilities entail opposite counterfactual conditionals.

We cannot answer this problem in the way we tried to answer the Bizet-Verdi case. First, as just said, although this case can be interpreted as an epistemic

---

compatriots, both would be Spanish because it would be easier for both to be Spanish than one of them to be French or the other to be Italian. This shows that the sense of asking such question is dubious. If our pets were of the same species, would I have a cat, or would you have a dog? God knows, but who cares.

[7] I thank Zoltan Gendler-Szabó for calling my attention to this important semantic fact.

uncertainty, our account must at least make room for interpreting it as a case of genuine metaphysical indeterminacy. Second, the principle of the conditional excluded middle seems clearly valid in this case, since – unless we grant the possibility that the coin lands on its edge – it would land either on its tails *or* on its heads, if tossed.

Nonetheless, it seems that our considerations in the previous section apply to this case as well. We cannot reasonably deny that the ascription of both abilities to the coin entails counterfactual conditionals. For I do not merely express an epistemic attitude about the expected result of the coin-tossing by saying that it *might* be heads. Thus, we need to explain how the opposite 'would-counterfactual' conditionals identifying possibilities can be both true of the coin at the same time.

Let us assume that tossing a coin is a genuinely indeterministic process: the coin can land heads up when it is tossed but, in the exact same circumstances, it can also land tails up. Indeterminism itself does not entail probabilities. It might not be determinate which result will occur without there being any probability that one will occur rather than the other. However, it seems that in this case there is also a certain probability with which the coin will land either heads or tails. If the coin is fair and if it is tossed fairly, we would ascribe 50 per cent probability to each possible outcome; that is, to the coin's landing on either of its sides. What we need to understand first is how the coin's relevant abilities can ground the two possibilities if the ascription of them entails counterfactuals with opposite consequents.

If abilities ground possibilities, the coin must have both the ability to land heads up and the ability to land tails up (that is to say, heads down). These abilities can be specified with the circumstances in which the coin is tossed. Given our account of abilities, it must be true of the coin that it would land heads up if it were tossed in circumstances C *and* it is also true of it that it would land tails up if it were tossed in the same circumstances. The two counterfactuals specify the coin's two different, though not distinct, abilities. But how can both conditionals be true? Can 'If C obtained at t_0, then φ would occur t_1' and 'If C obtained at t_0, then ψ would occur at t_1' both be true when φ-ing and ψ-ing at the same time are incompatible?

There is certainly no contradiction in the claim that something *can* φ and *can* also do something incompatible with its φ-ing. If, as I discussed earlier, one uses would-conditionals to express some sort of necessitation-relation like Lewis seems to do, it follows that the two counterfactuals cannot be both true. For the conditions specified in C cannot necessitate both φ and ψ. Using the standard possible world apparatus, one can argue that there is not a closest (set of) C-world(s) in which both φ occurs and ψ occurs.

However, as I argued, there is another interpretation about the use of would-conditionals, especially in the context of ability-ascription. If we use such conditionals in order to specify abilities which ground certain possibilities, then we must hold both conditionals true. For only the truth of *both* conditionals entails, correctly, that a fair coin has *both* of the abilities. And it must have both abilities since it is these abilities that ground the relevant possibilities necessary for genuine indeterminacy, which means that both results *can* occur in the same circumstances. How can we explain this?

First, the coin's dual ability is not to be confused with the idea that something can have the single 'conjunctive' ability to φ and to ψ (and hence to not-φ) at the same time. Ascribing that property would entail that something would both φ *and* fail to φ at t_1, if C obtained at t_0. This is indeed a contradiction, and hence nothing can have such ability.[8] But from the ascription of two distinct abilities it does not follow that an ability with such 'conjunctive manifestations' can be ascribed as well. The possibility that a coin lands heads up at t_1 and the possibility that it lands tails up at t_1 are logically compatible. But these two possibilities do not entail that it is possible that the coin land tails up *and* heads up at t_1.

Hence, the specific problem posed by the existence of genuinely indeterministic processes to our account of contingent possibilities is the following. Our aim is to specify different possibilities with the help of specifying objects' abilities which are their ground. But, further, we also want to specify the relevant possibility-grounding abilities with the help of counterfactual conditionals. In the case of coin-tossing this means that the coin must have the ability specified as 'The coin would land heads up at t_1, if it were tossed at t_0' and the ability specified as 'The coin would land tails up at t_1, if it were tossed at t_0'. But the truth of these two conditionals appears to be incompatible with each other.

Should then we specify indeterministic abilities by granting the truth of two apparently contradictory counterfactual conditionals? Or should we admit that indeterministic processes constitute a context in which our account does not apply? No, we should not. For on a more careful investigation of the relevant circumstances in which the respective abilities would be exercised the contradiction turns out to be merely apparent. There is a contradiction here

---

[8] This means that whatever role counterfactual conditionals like 'If Hobbes squared the circle, he would have been a famous mathematician' play in our language, they do not ascribe any ability to Hobbes.

only if we use a simple conditional analysis of abilities, rather than my revised formulation.

The circumstances in which we ascribe abilities that are *indeterministic* have the following important common characteristic: from the perspective of *one* of the respective abilities, the instantiation of the *other* ability by the same object should always be regarded *as part of the circumstances* in which the ability is to be exercised. What makes a process indeterministic is that that other ability 'competes', as it were, for its being exercised in the very same circumstances. Using our former terminology, abilities which characterize genuinely indeterministic processes are mutual intrinsic 'finks' for each other.

This means that the ascription of the two abilities of the coin in our example is not connected to *exactly* the same conditional. The ability to come up heads would be exercised only if the exercise of the ability to come up tails were 'trumped' and hence not exercised; while the ability of come up tails would be exercised only if the ability to come up heads were 'trumped' and hence not exercised. Alternatively, we may say that the ability to come up heads would be manifested only if the coin were tossed, *and* its ability were not finked by the coin's other ability, which is its ability to come up tails if tossed and vice versa.

As we have seen in the previous chapter, it is an important feature of many abilities that they can change after the time at which they begin to be exercised. Our schema says that an object *o* has the ability A at *t* if and only if an event of type Φ involving *o* would occur no later than *t'* if *o* were in circumstances of type C at *t* and it *did not change with respect to having or lacking A until t'*. A (fair) coin has the ability to land heads and the ability to land tails at t_0, that is, at the time when it is tossed. But it cannot *retain* both abilities until the time of its final landing. By some time before landing, it must lose one of its abilities. And what explains the coin's losing that ability is the interference of the other 'competing' ability.

The coin-tossing is, of course, just a special case. We can conceive objects with as many 'competing' abilities as we want. Each of these abilities 'competes' for being manifested in the sense that they would be manifested only if, during the process leading to their manifestation, they would 'fink' all the others; rather than being 'finked' by another. It is then true only in one sense that indeterministic abilities can be exercised in exactly the same circumstances. The circumstances in which they would be exercised are the same for *all* abilities. But they are not the same for *each* ability individually. This is because a process is genuinely indeterministic only if each of the abilities counts as a fink from the perspective of all others.

As mentioned, in the case of coin-tossing, and in the case of most nondeterministic processes,[9] we also assign *probabilities* to certain events. In our example, for instance, it is reasonable to assign a 50 per cent probability to the event that a (fair) coin will land heads up if tossed fairly, and a 50 per cent probability that it lands tails up if tossed in the same circumstances. How can we understand such probabilities?

On one interpretation, such probabilities are mind-dependent in the sense that, by assigning them, we project our evidence-based rational expectations, which can come in different degrees of strength measured by probabilities, onto the future nondeterministic events. Probabilities are objective then only in the sense that they must be based on rational evidence (hence they are not entirely *a priori* or 'personal'), but not in the stricter sense that they are mind-independent features of reality. Although I am sympathetic to this interpretation, I do not want to commit myself about it here.[10] So I need to address at least the following *conditional* question: if there are any mind-independent objective probabilities about particular events, how can we understand them?

According to one understanding, we may want to understand them as features of events. This would mean that a fair coin has the ability to land-heads-with-50-per-cent probability and it also has the ability to land-tails-with-50-per-cent probability. This interpretation has the advantage that it straightforwardly explains how we can ascribe two different abilities to an object, even if the consequents of the conditionals entailed by their ascription are detached. For the join exertion of these two abilities involves no contradiction. The event landing-heads-with-50-per-cent probability and the event landing-tails-with-50-per-cent probability can, after all, occur at the same time.

However, this strikes me as an utterly bizarre view. First, in order to make sense of it, we should introduce some sorts of entities like 'landings-tails-with-50-per-cent probability' and 'landings-heads-with-50-per-cent probability', *both of which* actually occur when a coin lands tails up and when it lands heads up. After all, a coin's not landing tails up is perfectly compatible with its landing-tails-with-50-per-cent probability. In fact, it would not even be right to describe what has happened as its landing-heads-up, only as its landing-heads-with-50-per-cent probability. In general, it would be impossible to decide which ability's exercise an event might *manifest*.

---

[9] In fact, some may hold that in the case of all indeterministic processes, except perhaps free human actions, there *must* be some prior probability of which abilities will be exercised. But this is a complex issue and I do not want to commit myself about it here.

[10] For an interpretation of probability which seems to be close to this see Hoefer (2007).

For this reason, it is more plausible to understand mind-independent objective probabilities as the characteristics of objects' (or systems') abilities rather than characteristics of the events which manifest them. The 50 per cent probability that a (fair) coin lands heads should mean then that there is a 50 per cent probability that the coin's ability to land heads is exercised if it is tossed fairly; and there is a 50 per cent probability that its ability to land tails is exercised when it is tossed fairly. The coin can have both abilities before and when it is tossed. But only one of them can be exercised by the time it lands. Probabilities are then, fundamentally, the probabilities of which abilities of objects (or systems of objects) are going to be exercised in the exact same circumstances. The probabilities of the events that occur are derivate of the probabilities attached to abilities.[11]

Finally, we also need an explanation why the probability of heads and the probability of tails cannot exceed 1, why the probability of heads and tails are 0 and why it is the case that if the probability of heads is 50 per cent, then the probability of tails must be 50 per cent too. What explains this is that if the coin were tossed in circumstances C, (a) it would exercise either its ability to land heads or its ability to land tails, but (b) it is logically impossible to exercise both abilities together, and (c) there is not any other ability which the object would exercise in the exact same circumstances C, and which would be logically incompatible with the exercise of the other two abilities.

In the case of the coin-tossing, the satisfaction of the last condition is guaranteed by logic: if we consider the coin as – or as if it were – a two-dimensional object, it *can* land only on one of its sides (the possibilities are exclusive), and it *must* land on one of its sides (the possibilities are exhaustive). And even in the more complex cases, the probability that two or more probabilistic abilities cannot be exercised together, and hence the probability of their joint exercise must be zero, is guaranteed by logic. For it is a matter of logic that nothing can exercise the ability to φ and the ability to not φ at the same time, in the same circumstances.

But in many cases, whether or not we have managed to identify an exhaustive list of abilities, neither of which can be exercised at the time when the other is, is an empirical matter. Suppose we break some windows made of the same kind of glass, and we observe that, in the exact same type of circumstances (when the same forces are exerted on them by the same object from the same direction), some of them break into two pieces, while others do not. Can we be certain that

---

[11] I shall come back to the significance of some of these problems in the next chapter when I will discuss in more detail the relation between manifestations and events as the results of exercising an ability.

the probability that the ability to break into two pieces *or* the ability to resist disintegration will be exercised with probability 1? Not necessarily. Since it is *logically* possible that the same kind of glass in the exact same circumstances might break into more than two pieces. The question whether or not our list is exhaustive can depend only on logic: since nothing can break and fail to break at the same time, the two abilities cannot be both exercised in exactly the same circumstances. But if we are also interested in some more specific probabilistic abilities – as the ability to break into two, three and so on pieces – whether or not our list is exhaustive is always an empirical matter.

In sum, the ability-based account seems to me the most promising interpretation of objective, single case probabilities in the world. However, we cannot be certain that there are indeed such probabilities; or that there is any objective indeterminacy in the world. Perhaps, there are not any 'genuinely' nondeterministic processes because the circumstances in which an event φ would occur and in which φ would not occur are never *exactly* the same. If this were right, then we would assign probabilities to events only because we cannot explicitly specify the circumstances in which they would occur.[12]

Abilities might be either deterministic or nondeterministic in the sense explained earlier. But can objects also have probabilistic *dispositions*? It might seem that they can *only if* the probabilities that one of the object's (relevant) abilities is going to be exercised when certain conditions obtain are *not* (nearly) equal. A fair coin is able to land heads up and is also able to land tails up when it is tossed, but it is not disposed to do either. Dispositions ground tendencies, and if there is a 50 per cent chance that the coin will land tails up when tossed, it has no *tendency* to land heads up when tossed.

However, if the probability that one of the abilities are going to be exercised is more than 50 per cent, then it seems plausible to say that an object o is disposed to φ in C. For assigning that probability entails that the kind of

---

[12] However, my account about how we specify abilities through their connection to counterfactual conditionals does have an interesting consequence about determinism and probabilities. Abilities at different levels of specificity can be identified with the help of the inclusions of different factors in the antecedent of the conditional that the ascription of abilities entail. Suppose we specify an ability $G^{\lambda 1}$ with a conditional the antecedent of which include factors $\{c_1 \ldots c_{n-1}\}$. $G^{\lambda 1}$ is nondeterministic if we can also specify another ability $H^{\lambda 1}$ with a counterfactual that has the same antecedents $\{c_1 \ldots c_{n-1}\}$, but with a consequent that is incompatible with the consequent of the conditional specifying $G^{\lambda 1}$. We might even assign some probability with which the ability would be exercised if $\{c_1 \ldots c_{n-1}\}$ obtained. Nonetheless, a more specific ability $G^{\lambda 2}$, which is specified by a conditional with the antecedent $\{c_1 \ldots c_{n-1}, c_n\}$ might be deterministic, because the object does not have the corresponding, more specific ability $H^{\lambda 2}$. What this indicates is that the occurrence of the same event can manifest both a nondeterministic and a deterministic ability. This has important consequences in the context of debates over 'deterministic chance'. About the latter, see Mellor (2000b), Hoefer (2007), Glynn (2010) and Eagle (2011).

objects which *o* exemplifies would φ more often than not if C obtained. In fact, almost all dispositions are 'probabilistic' in this sense since *o*'s having the disposition to φ in C entails only a tendency that objects of kind O φ in C. However, this does not mean that there must be a single-case objective probability of *o*'s φ-ing in C on a particular occasion. Dispositions, as I have argued, ground the truth of *ceteris paribus* generalizations, but they need not ground any probability.

There are many other difficult and interesting philosophical questions about the interpretation of probability, but this is certainly not the proper place to discuss them. Important here is only that (a) my proposed account of abilities can easily explain the nature of indeterministic processes; and that (b) if there are objective *a posteriori* probabilities, then they are grounded in objects' nondeterministic abilities. All this is perfectly compatible with the view that these abilities are specified through their logical connection to certain would-conditionals. And further, it is also compatible with, though does not require, the view that if an object has the disposition to φ in C, then the probability that it will exercise its ability to φ in C is significantly higher than the probability that it will not φ in the same circumstances.

Ascribing abilities and reasoning with possibilities

As we saw in the previous chapter, the more generic the ability we want to ascribe is, and hence the more abstract possibilities are we interested in, the more conditions we need to *explicitly* specify in the antecedent of the relevant counterfactual conditional. For the more generic a modal property is, the less and less conditions need to be *actually* satisfied in order to correctly ascribe it to an object. The ascription of more specific modal properties requires less specific conditional antecedents, since when we ascribe them to an object at *t*, we *presuppose* that in the particular situation in which we do so the unmentioned conditions are all actually satisfied.

This is certainly no news. Suppose a particular vase has the ability at t_0 to break. According to TCA this entails that the vase would break before t_1 if it were dropped at t_0. But it would not break, if it landed on some sufficiently soft material, or if it were dropped from a height of 2 centimetres, or if it were dropped on the moon, and so on. Conversely, a hard piece of concrete which is able to retain its integrity even when some strong force is exercised on it would break if it were dropped from a height of 200 metres on a granite's edge.

Everyone agrees that the conditionals with which we *initially* specify an ability with reference to a salient condition (e.g. dropping) are incomplete. The initially mentioned conditions are also often too specific for most purposes. In a more precise formulation, instead of dropping, we should talk about exerting a kind of force from a certain direction on the object. Dropping, or hitting an object with a hammer, just describes specific ways to exert such forces. More interestingly, instead of breaking, we should talk about splintering or fracturing, since breaking, as I shall argue in the next chapter, presupposes that a (certain degree of) *fragility* is manifested.

Notice, however, that by introducing this more precise language, we leave behind the original idea that the meaning of modal – especially dispositional – terms should be given in term of what makes them *manifest* in the original sense of manifestation, in which it means a directly, sensibly observable event. Fracturing might be directly observable, but exerting a certain degree of force from certain direction need not be. I shall return to the significance of this, again, in the next chapter.

What the phenomena of masks and antidotes illustrate about the nature of modal properties is that, when we ascribe some generic ability, the antecedent of the corresponding conditional must also include suppositions about *inhibitory* factors. If a mug is actually filled with protective material, then it does not have the specific (in the actual circumstances) ability to break when dropped. Nonetheless, there is certainly a sense in which it is able to break, even if this ability grounds a more abstract possibility. It is able to break, because it has the more generic ability specified by the counterfactual conditional 'If the mug were dropped and were not filled with protective material, then it would break'. Even *the actual non-breaking of the mug* when it is dropped is compatible with having this more generic ability.

Thus, it is obvious that there will be cases in which the counterfactual conditional with which we initially aim to identify a modal property of an object is underspecified, even if it is true of a particular object's maximally specific ability in a given situation at *t*. We cannot identify some *more generic* modal property of an object without explicitly specifying more conditions that need to obtain for its exercise. This is a natural consequence of the generality of the modal property we want to specify, and it does not undermine the importance of the conditional analysis as a conceptual means to specify them.

Take any ordinary property and ask someone to provide necessary and sufficient conditions for its being instantiated. It's a safe bet that some conditions will be missing so that someone can come up with examples where something or someone does not have the property even if the already specified conditions of its possession are all satisfied. You say that a bachelor is an unmarried male. But then you will be reminded that your ten-year-old son is not a bachelor. Fair enough, you need to mention the condition that unmarried males older than, let us say, twenty years are bachelors. But then you are warned that the pope is certainly older than twenty years, but he is not a bachelor. Then you say that a bachelor is an unmarried adult male unless he is . . . Who knows where this process ends? But this does not make your initial characterization useless or vacuous. Because when you ask whether or not a particular person, let us say Jack, is a bachelor, you can correctly say that yes, he is *because* you know that he has never been married.

Therefore, the interesting question is not whether very simple conditionals with a single antecedent are sufficient for specifying all generic modal properties, for it is evident that they are not. Practically, in order to initially identify a modal property, we need to identify *some salient aspects* of the circumstances in which it *would* be exercised and leave unmentioned many others. When we apply our analysis to a particular case, we presuppose that many, yet explicitly unspecified, conditions *are* actually satisfied. If they are not, the object does not have the specific ability there and then. But we might be interested in whether it has a more generic ability, because we want to identify some more abstract possibility. Then we try to enrich the conditional's antecedent with more explicitly mentioned conditions.

Here is an example to illustrate the idea. Suppose that, while driving on a highway, I wonder whether my car can – that is, has the ability to – take me to a certain destination within two hours. Initially I assume that it does. When I hear a message on the radio that there is an accident on the highway, I learn that it cannot. I know that my car would take me to my destination within two hours if I kept my average speed. But I also know that it would take me to my destination only if I kept a certain speed and if there were no accident on the road. When I ascribe the ability which grounds certain possibilities, I presuppose that many conditions *are* satisfied. When I learn that they are not satisfied, I know that my car (and, derivatively in this case, me) does (do) not have the specific ability which I previously thought we have in the particular circumstances. But this does not mean that I must have lost the more generic ability; it is only that *that* is not my concern anymore.

However, more generic abilities often *are* my concern. Suppose I realize that I cannot reach my destination within two hours because my car will soon run out of fuel. Thus, given my actual situation, I do not have the specific ability to arrive at my destination in due time. But I know that I have the more generic ability that if I continued to drive with the certain speed *and* my car had enough fuel in its tank, then I would reach my destination in time. Thus, I try to check whether there is a petrol station on the highway where I would stop to buy fuel. Ascribing a more generic ability can motivate us to actualize certain conditions and then to acquire specific abilities that we have previously lacked.[13]

Thus, the ascription of abilities at different levels of generality plays a crucial role in ordinary practical reasoning and complex planning. Counterfactual conditionals provide the link between the ascription of abilities on the one hand, and our theoretical and practical reasoning about, and with, them on the other. It is with the help of such conditionals that we initially identify an ability, and it is by enriching their antecedent that we can identify more generic abilities.

The more generic the ability we aim to ascribe is, the more detailed knowledge of the antecedent of the relevant conditional we need to have. We aim to identify more generic abilities, because this may increase our knowledge about whether an object has the more specific one in a concrete situation, but also, because it helps us to identify which environment we need to find or create in order to realize a possibility.

Interactions and the limits of generality

It is often noted that modal properties are not metaphysically independent of each other. As already mentioned and discussed in Chapter 3, most natural (or physical) abilities can be instantiated only if some objects possess some other abilities which are their 'partners'.[14] Some solids are soluble, and some liquids are solvent. The exercise of these chemical substances' abilities is mutually conditioned on the presence of the other.

However, these are only the simplest cases of 'partnership'. Four hydrogen atoms and a carbon atom are able to compose a methane molecule. But they can do so only together: each hydrogen atom needs the other three hydrogen atoms and the carbon atom to exercise this ability. And the more complex (chemical)

---

[13] Another, more dramatic, example for those, who are familiar with the story, is the case of *Apollo 13*. Initially, right after the spacecraft was seriously harmed, the possibility to bring it back to the Earth was only very abstract and the ability of the spacecraft to return was very generic, or 'iffy'. But as the astronauts managed to make more and more conditions actually satisfied eventually the ability became more specific and the possibility more concrete (so much so that finally it has been realized).
[14] The concept of 'disposition partners' was introduced by C. B. Martin and used later by Lewis (1999/1997) and Heil (2003).

phenomena we investigate, the more complex the systems of such 'partnership' will become.

Masks and antidotes are 'bad partners' with those abilities which they mask or to which they are antidotes in the sense that they can – or have the tendency to – *prevent* or *inhibit* the manifestation or exercise of another dynamical-modal property.[15] Yet, an object and substance with such modal properties can function as masks or antidotes only if another object or substance has an ability (or, if you like, *l*iability) *to be masked or antidoted* by them. Interdependence of modal properties is the rule, not an exception. Almost every disposition is manifested, and every ability is exercised, in some environment in which some other *contributory* abilities are present and some *inhibitory* abilities are absent.

The proper operation of human immune system masks the manifestation of certain viruses' and bacteria's ability to cause certain forms of illness. Unfortunately – in this case – the presence of yet another type of virus can contribute to the manifestation of the originally masked abilities. HIV infection *enables* certain viruses and bacteria to exercise their ability to cause certain kinds of disease by *disabling* the infected person's immune system, which normally prevents them to do so.

Does this interdependence of modal properties pose an unanswerable challenge for the conditional analysis? Not necessarily. If the purpose of a philosophical analysis is to explain the content of certain dispositions and abilities without any reference to some other, then the phenomenon of masks and antidotes does indeed seem to give rise an unanswerable difficulty. But if our purpose is to identify an ability or a disposition by specifying it with the help of counterfactual conditionals, then reference to other modal properties in the analysis is rather natural. Our RCA[A] does not aim to reductively analyse abilities, but it helps specifying them, at least partly, with reference to the presence or absence of other modal properties in the circumstances in which an ability would be exercised.

By using the RCA[A] schema, we can identify modal properties at various levels of generality. Which specific factors or conditions c_s, $c_1 \ldots c_n$ need to be included in the antecedent of the conditional *depends on how generic* the ability that we aim to specify is. When specifying a maximally specific ability we need to explicitly include only a single factor c_s – often called the 'stimulus event' – in the antecedent of the conditional. In case of maximally specific abilities, as we have seen, the *presence* of the masks and antidotes or the *absence* of some

---

[15] As Fara observes in Fara (2005).

necessary contributory abilities in the actual circumstances at t (when the property is ascribed or supposed to be possessed by the object) is incompatible with having the ability. Hence when we ascribe such abilities to objects, we simply presuppose that, in the actual circumstances at t, masks and antidotes are absent and necessary contributory abilities are present.

Most maximally specific abilities are extrinsic to a high degree, since their possession is very sensitive to the alteration of their bearer's circumstances. But as we have seen earlier, our aim is often to identify some less specific and hence more intrinsic ability, the possession of which is less sensitive to the variations of extrinsic circumstances. The RCA^A provides a general schema for specifying abilities at various levels of generality which ground those more abstract possibilities. We can use the schema provided by RCA^A in order to specify more and less specific dispositions and abilities through their logical connection to counterfactual conditionals.

When we ascribe specific abilities to an object, we might include in the antecedent of the conditional only a single factor c_s because we assume that all contributory properties are present, and all inhibitory factors are absent, *in the concrete situation* in which the ability is ascribed. If they are not, the conditional is false – but so is the statement which ascribes the (specific) ability to the object. As we ascribe more and more generic abilities, we need to explicitly include further conditions in the antecedent. We need to mention them because our aim then is to ground some more abstract possibility which can exist *irrespective of* whether or not some of certain conditions are actually satisfied in the situation in which we ascribe an ability.

However, it seems that we cannot explicitly mention the absence of every conceivable counteracting factor in the conditional's antecedent.[16] Suppose that, initially, we identify a relatively less generic ability with the help of a conditional which explicitly specifies certain conditions of its exercise. When we add further conditions to the antecedent of the conditional, we identify more generic abilities; and gain more knowledge about which conditions need to be satisfied for the ability to be exercised and for a possibility to be actualized.

But there is a limit to generality. Since the relevant conditions must include the absences of inhibitory factors like masks and antidotes, it seems that we can never explicitly specify a maximally generic ability. For *it is always conceivable* that there might exist some further inhibitory factors, the presence of which would thwart the exercise of the ability we aim to specify.

---

[16] See Bird (1998, 231).

Some philosophers consider this as an unanswerable objection to the conditional analysis. Their worry is that when we say that *all* potentially interfering factors must be absent otherwise the ability would not be exercised, we seem to be saying something totally empty. This may be one of the reasons why we do not mention such factors when we initially try to identify a dynamic modal property. Some fragile objects certainly break sometimes, so at least at those times all counteracting factors must be absent. But there is no way to identify a generic ability with the help of a conditional which can explicitly include the absence of *all* sorts of counteracting factors in its antecedent.

However, contrary to the opinion of such critics, this does not make the analysis vacuous. For a philosophical analysis – as Hegel once remarked – should not aim to formulate a *single* definition.[17] Rather, it can consist of a *logical series* of definitions (which may or may not reflect the historical evolution of some concept). An initial definition is only a start of a *process* of specifications (where 'process', again, means a logical, not necessarily a temporal, sequence). As we include more and more conditions in the antecedent of a conditional, we can (*i*) identify a more general modal property to φ; which (*ii*) grounds a more abstract possibility to φ; that (*iii*) exists irrespective of whether the object to which we ascribe the property at a time possesses or lacks the specific ability to φ then and there.

Here is a famous historical example to illustrate such a process.[18] It has been discovered that penicillin, when administered in an appropriate dose, can – and hence has the ability to – cure streptococcal infections. But it has also been found that some patients have not been cured from the infection even when they received the treatment that was sufficient for the recovery in many other cases. Further medical research needed to identify the conditions in which treatment with penicillin fails to cure an infected patient.

Most of these conditions include some interfering factors, for instance, the previous use of antibiotics, or the joint application of other medications. No one would think that the need for a further process for specifying those conditions renders the initial counterfactual – 'if penicillin were administered to a patient with a sort of bacterial infection in a certain dose, she would be cured from streptococcal infection' – was useless or even generally false. No one would think that penicillin, when administered in appropriate dose in proper circumstances,

---

[17] See Hegel (1812/2010), 'Introduction'. Whatever analytic philosophers think of Hegel's philosophy in general, he seems to me right in that doing philosophy often involves a *logical process* rather than providing single definitions. This is one of the most important messages of his *The Science of Logic*.

[18] The example is borrowed from Hempel (1965, 394).

does not have the ability to help people recover from certain types of disease just because it is always conceivable that some new counteracting factor (e.g. the parallel use of another kind of medication) can undermine its effect.

Penicillin has the generic ability – and even the disposition – to cure people at a certain age from certain kinds of illness, if it is administered in the appropriate dose *in the absence of* certain intervening factors $f_1 \ldots f_n$. Its most abstract, most intrinsic, ability is identified by the counterfactual 'If penicillin would be administered in a given dose at time *t* to a patient suffering bacterial infection of type *I* below at the age *x* and in the absence of certain intervening factors $f_1 \ldots f_n$, the patient would recover by time *t*'. Doctors decide whether to prescribe penicillin in a concrete case – that is to say, whether penicillin has the *specific* ability to cure *this* particular patient – by learning more about its *modal properties* and the patient's *actual conditions*.

Thus, our practice to identify abilities or dispositions with the help of counterfactual conditionals is directly connected to how we reason both theoretically and practically. Whenever penicillin is administered without the desired effect, we do not reject the initial counterfactual entailed by its having a certain modal property. Rather, we become theoretically motivated to learn more about penicillin's more generic ability to cure certain diseases. We try to identify more and more conditions which explain why our former ascriptions of the specific ability in some concrete situations turned out to be mistaken. By doing so we also try to identify a more generic ability that penicillin possesses even when it has turned out not to have the more specific one.

In our example about penicillin's ability, it is obvious that we cannot identify *all possible* intervening factors. Any unfortunate patient can, for instance, die for infinitely many different reasons before penicillin can cure her. Consequently, we must always leave some possible interfering factor 'blank', and so we cannot explicitly identify penicillin's *most generic* ability. More generally, if the exercise of an ability is not instantaneous – as it is not in the case of recovery – then it is always conceivable that a process which started in certain conditions will not be completed and the event which manifests the exercise of the ability will eventually not occur.

Since abilities ground *contingent* possibilities – rather than possibilities that follow by necessity from some logical or mathematical truth – the possibilities that they ground can be more or less abstract. Administering *this* dose of penicillin can cure *this* patient *this* time. But even if it cannot, it can cure *some* patients at some other time. Whether we are interested in the former or the latter ability depends on our theoretical and practical interest. But that these more or less abstract possibilities exist does not depend on our knowledge and purposes. We discover such possibilities; we do not postulate them on some *a priori* ground.

Sometimes we want to understand why something happened at a particular time in specific type of circumstances, or what could have happened then and there. In such cases, we are interested in very specific abilities. But sciences of the more theoretical kind aim to identify more abstract abilities that can explain the occurrence of some very generic type of events (e.g. acceleration); or they want to design devices optimal to satisfy certain ends (to reach the Mars). For such theoretical and practical purposes, we need to identify more generic abilities and dispositions, because we want to know what would happen in various counterfactual circumstances. Then we need to include as many contributory and inhibitory factors in the antecedent of the conditional specifying an ability as we can discover.

But again, there is a limit. We cannot identify objects' and substances' *maximally generic* modal properties. A maximally generic modal property should be specified with the help of a counterfactual conditional the antecedent of which includes the condition of absence of every conceivable counteracting factor; and there does not seem to be any *a priori* limit to the number of such factors. Yet, by using counterfactual conditionals, we can identify modal properties at a sufficient level of generality so that we can satisfy our theoretical and practical purposes.

It is true that when we say that a person, object or substance can φ, we may ascribe a very generic ability to them. But we can ascribe such very generic capacities to φ *simpliciter* only because they have more specific abilities to ψ in C. An object could have the (generic) capacity to break only if it would break if certain specific and specifiable conditions were obtained. In fact, ascribing such generic capacities is not practically or theoretically very useful. To ascribe *a* capacity to break does not help much in most contexts of reasoning, since virtually every macroscopic solid *can* break. Reasoning with abilities starts when we are interested in specifying the conditions in which they *would*.

Modal properties without conditionals

Consequently, even if it is true that we cannot specify maximally generic abilities, this does not undermine the theoretical significance of the conditional analysis of abilities. It only sets a natural limit on what we can know about possibilities. But some philosophers argue that there are important theoretical properties that are not logically connected to any conditional, even if they are modal. Such modal properties, they argue, are either (*i*) 'continuously exercised' or (*ii*) may be exercised in any arbitrary selected conditions whatsoever.

My initial response to this objection, which I emphasized already in the first chapter, is that not all modal properties are of the same type. It might be the case that there are certain types of modal properties that can be identified only with reference to the type of events that occurs as the result of their exercise. This does not undermine the significance of the conditional analysis of abilities. For a modal property that is continuously exercised can hardly identify any possibility in our world. Rather, it used to describe its actual operation.

It is less clear than it is often assumed, however, that such properties indeed exist. Here is an example to explain why. George Molnar mentions rest mass in General Theory of Relativity as an example of 'continuously manifested' powers. In this example, it is assumed that an object's rest mass is 'continuously manifested' in its interaction with spacetime (Molnar 2003, 86–7). Molnar admits that this example is contentious.[19] But he claims that nothing hangs on the examples. It is enough to admit that such powers (i.e. such type of modal properties) are conceivable.

I am not sure they are. For if a Molnar-type 'power' is continuously and always exercised, and hence it is necessarily manifested whenever it is instantiated, then it is hard to see what distinguishes a modal *property* from a mere description of an actual event or process. A body having a rest mass certainly has many (nomic) abilities and dispositions related to how it can (or tend to) physically interact with other bodies having some other physical properties. But the continuous interaction of a body's mass with spacetime is the description of an actual process, not a 'manifestation of a continuously exercised ability'.

But cannot an ability be manifested continuously, but only accidentally?[20] I can see only two ways in which this can happen. (*i*) Perhaps there are some circumstances on the occurrence of which its exercise depends, and in the absence of which it would cease to be manifested. But then the manifestation of the ability must be conditional on the continuing presence of something. (*ii*) Alternatively, we have to imagine a case when something happens continuously but entirely accidentally, non-nomologically, and without any possible condition the presence or absence of which could stop it happening.

---

[19] Moreover, it is not clear to me why the relation between rest mass and the curvature of spacetime in GRA is an 'interaction' rather than a theoretical identity; but that is a complex issue that we need not pursue here.
[20] Thanks to a referee for raising this issue.

The latter strikes me as being very near to an inconceivable scenario. But if someone else finds it conceivable, then I am ready to admit that this case does indeed provide an example for unconditional abilities. It does not matter much though, since it follows from the nature of the case that the ascription of such abilities can play no role whatsoever in our reasoning about possibilities. I understood accidents earlier as exceptions to regularities that do require explanation. 'Accidentally continuously manifesting abilities' would, however, be occurrences which, by their very nature, unexplainable. Hence, if such things exist, they must be beyond our power to understand them.

It might be objected further that – as we have seen in Chapter 3 – certain dispositions are manifested by objects' or substances' having certain features rather than by the occurrence of some type of event. Tigers are genetically disposed to have four legs. This disposition is not manifested only when some specific condition obtains at a time. Can we not understand such dispositions as 'continuously manifested' modal properties that are 'unconditional'? After all, it seems that the identification of such dispositions requires only the specification of the feature that manifests them. Tigers are disposed to have four legs; spiders are disposed to have eight legs. We need not say more in order to specify these dispositions.

But then a natural question arises about why we should regard such features as manifestations of some dispositions in the first place. The answer is that, while nothing can have a determinate rest mass without interacting with spacetime in a determinate way, something can – as we saw in Chapter 3 – be a tiger without having four legs and can be a spider without having eight legs. Tigers are mammals, spiders are Arthropoda, with no exception; hence these are not abilities or dispositional properties of their bearers. But having a certain number of legs is the manifestation of a disposition because tigers are genetically disposed to have four legs and spiders are genetically disposed to have eight, even if some of them happen to have less.

Phenotypic and species-specific behavioural features are dispositions to the extent that they are conditioned on the type of genetic processes which can be 'masked', or which can have 'Achilles heels'. A tiger is genetically disposed to hunt. It is genetically disposed to hunt, even when he is unable to hunt because he lives in a zoo. And he is genetically disposed to have four legs, even if he lost one in a fight.

For the purposes of scientific classification, many generic nomic dispositions are identified exclusively in terms of the types of events or occurrent features which manifest them. And many such dispositions, unlike teleological dispositions, are not manifested by intermittent events – like hunting – but by features that an object can continuously have. It is easy to see why being a predator is a dispositional property. A tiger is disposed to hunt even in moments when he is not hunting. But most tigers have four legs all along their life. Hence it is not so conspicuous that having them is a manifestation of a disposition just as hunting is.

Nonetheless, having four legs is conditional on how the tiger's genes transmit genetic information. The number of limbs, as well as other kind-specific traits, depends on how an organism's genes operate during a certain period of its life. Species-specific phenotypic traits do not appear unconditionally. All of them are conditional on the organism's genes and the (intrinsic and extrinsic) environment in which the trait or feature develops.

Another sort of examples about the allegedly 'unconditional' modal properties are the so-called spontaneous powers. Some nomological processes, like some forms of radiation or the emission of certain kind of particles, are 'spontaneous' in the sense that, although they are not manifested 'continuously', they may or may not occur in any condition.

I do not deny the phenomenon, of course. However, as before, it is unclear why spontaneous radiation is a property at all rather than being the description of a sort of events. Of course, those events occur, as a result of the exercise of some particles' abilities and dispositions. But the ascription of those abilities and dispositions, contrary to Molnar's claim, is logically connected to the truth of certain conditionals.

First, decay can happen as a result of particle interaction. So, the ability of *spontaneous* decay is manifested only in the conditions in which certain type of interactions are *absent* (e.g. the nucleus is not bombarded by other particles). It is that condition after all that makes decay 'spontaneous'. If that condition is not satisfied, a specific decay phenomenon could only be a 'mimic' of spontaneous decay (especially because decay in case of those interactions is also probabilistic).

Second, and more importantly, if we understand the phenomenon of radioactive decay as the exercise of an ability of an atom, then that ability is specified with the help of a conditional. The conditional says, very roughly, that if the nucleus of an atom of type U were in an unstable energetic state S at t and there were no other intervening factors present (as before), then it would emit a particle (or energy) of type P before time t' with probability x. It does not really matter that, by definition, as long as an atom is of type U, its nucleus is always in an unstable state, since this is just a matter of scientific nomenclature.[21] For

---

[21] As Hugh Mellor observes, 'Nuclear structure *is* changeable: the radioactive disintegration theory postulates that the same atom is transformed, when decay occurs, from one structure to another.'

the purposes of physical description (and the notion of decay makes sense only in the context of that description) what matters is the type of state in which the atom/nucleus happens to be at a time.

Finally, some allegedly 'unconditional' modal properties concern the capacities of human agents. An often-mentioned example is the 'spontaneous power' of human choice. The nature of human ability to makes choices is a topic of its own right, which I cannot discuss here in sufficient detail. Since deliberation and making conscious choices are certainly not deterministic processes – to the extent it is proper to call them processes at all – they must involve the exercise of some nondeterministic abilities. But the ability to make choices is not spontaneous *in the sense* that its exercise would not be conditioned on anything. Hence, comparing to the problem of choice to the case of radioactive decay as some do is rather misleading.[22]

Deliberation is a nondeterministic process in the sense that, whenever agents make conscious choices, it is not determined whether they will exercise their ability to choose to perform an action or their ability to choose an alternative in the same circumstances. However, unlike in the case of most physical abilities, there might not even be prior probabilities about which specific choice an agent is going to make. But this does not mean that choice-making is 'unconditional'. For there must be specific conditions in which an agent's ability to make choices is exercised.

Typically, (conscious) choices are made in conditions in which an agent has a reason to consciously exercise one of her abilities that are involved in making a choice. It is true that deliberation and choice are nondeterministic processes because whenever agents deliberate, they have both the ability to choose one option *and* the ability to choose another in the same circumstances (with the caveat that, although the circumstances are the same for all abilities, the condition of exercising each is not the same; exactly like in the case of nondeterministic physical processes that we discussed earlier). But this should not be confused with 'spontaneity', if 'spontaneity' means – as it does in the present context – that the exercise of the ability to choose is not conditional on anything. For our ability to deliberate and make choices is meant to enhance our rational control over our own behaviour, and if the exercise of abilities to make choices were

---

And 'a numerically identical atom undergoes over a period of time a series of drastic changes in its propensity to decay', Mellor (1971, 99–100).

[22] As, for instance, Lowe does in Lowe (2009, 149–50, 156). For more on the nature of abilities and of the indeterministic process that they involve, see Huoranszki (2011, 47–53; 2017).

unconditional, then the ability to choose could only undermine, rather than enhance, rational control.[23]

Relatedly, it is sometimes claimed that habits like being a smoker or being a (regular) swimmer are agents' dispositions, but their exercise is not conditioned on anything. Perhaps some smokers' smoking is conditional on being nervous or on being deprived of nicotine. But one can be a smoker even if one's smoking is not conditional on such circumstances. And, in the case of being a regular swimmer, it seems even more plausible to claim that the disposition is manifested as a response to some 'external stimulus'.

I agree with this, but all this does not mean that habitual behaviour is not conditioned on anything. Unless the behaviour is unvoluntary – in which case we would not call it a habit – the performance of the relevant kind of behaviour is conditional on the agent's choice (or at least on her intention). If it were not, agents would lose conscious control over their own behaviour whenever they perform a habitual action. But in the cases when they do lose conscious control over their own behaviour, their actions cannot manifest their habitual dispositions either.

It also needs to be mentioned that sometimes we describe agents' know-hows or skills by simply mentioning what they can do. It might be claimed that such 'can'-statements do not entail any conditional. An agent can add two large numbers, speak Spanish or swim in the sense that she knows how to add two large numbers, how to speak Spanish or how to swim. The acquisition of know-hows and skills typically requires learning and practice, so it is rare that one never exercises them.[24] But they might remain unexercised after they have been acquired. So, they might seem to be genuine modal properties which can be ascribed without the use of any conditional.

The nature of know-hows and skills is a philosophical topic of its own, which I cannot discuss here in any detail. Know-hows and skills seem to be neither dispositions nor abilities. Knowing how to φ does not entail any tendency to φ, so knowing-how is certainly not a disposition. And know-how is certainly not some *specific* ability either, since knowing how to φ at t does not entail that one is able to φ at t. One can know how to swim at t. Yet, one might not be able to swim at t if one's arms are broken.

Nevertheless, it seems to me that, to the extent know-hows and skills *are* described with the help of some 'can-statements', they are also connected to the

---

[23] See again Huoranszki (2011, 45–50).
[24] Though not impossible. An engineer can know how to manufacture a time-bomb; nonetheless, he might never do.

truth of some conditional. For we do not ascribe a know-how or skill to an agent unless we hold that she would φ in at least *some* appropriate circumstances, if she *tried*. What makes know-hows and skills special is that they also seem to be essentially connected to the normative evaluability of an agent's behaviour. Typically, even if an agent has the skill, or knows how, to φ, occasionally she fails to φ even when she tries; and when she succeeds, she can φ poorly or excellently. Hence, when the exercise of an ability presupposes know-how or skill, it seems almost always appropriate to ask whether, and how well, the agent φ-ed when she tried to φ.

6

Manifestations, events and causes

Becoming is a ceaseless unrest that collapses into a quiescent result.
(Hegel *The Science of Logic*, 21.93)

The main topic of this chapter is the conceptual and metaphysical relation between those events the occurrence of which can manifest or reveal the exercise of modal properties and these properties themselves. As we have seen in Chapter 4, one standard objection against the conditional analysis is that the occurrence of certain type of events may not manifest an ability, but only 'mimic' its exercise. In this chapter I shall argue that, rather than being an argument against the conditional analysis, understanding the phenomenon of 'mimicking' can help us to understand the very nature of events as a metaphysical category.

I shall also argue that the phenomenon of 'mimics' does not undermine the importance of conditional analysis. As we have seen, critics of the analysis often assume that objects' or substances' abilities and dispositions can simply be identified by their 'intentional object'; that is to say, by the types of events which would manifest their possession. The phenomenon of mimicking clearly proves, however, that this is impossible. The occurrence of a type of event that literally *manifests* a modal property is clearly insufficient for properly distinguishing and identifying such properties.

In fact, I shall argue that the occurrence of those events that have traditionally been called 'manifestations' directly reveals only something about objects' and substances' abilities to produce qualitative changes which are observable *for* us. But such changes themselves do not reveal the real nature of properties exercised. Consequently, the phenomenon of mimics does not only raise a technical problem concerning the connection between the ascription of abilities and the truth of counterfactual conditionals but also provides a further insight about the ontological role of modal properties. Modal properties do not only ground variably abstract possibilities, but they also explain what happens in the concrete reality.

It is almost universally assumed that the identity of modal properties depends, at least partly, on the states or events that would manifest the possibilities which they ground. As we shall see, however, the connection between modal properties and their manifestations is far more complex. I shall argue in this chapter that we can identify different kinds of events only by specifying the possibilities which they realize. But we can identify which possibilities an event realizes only by specifying the ability which is its ground.

It is obvious that modal properties are not ontologically and/or conceptually independent of the events or states that would realize the possibilities which they ground. But admitting this does not settle the important issue of priority. It is a further question whether the identity of abilities or dispositions is ontologically determined by the events that would happen when they become manifest; or, as I shall argue, what exactly happens in the world is ontologically determined by which abilities are exercised.

Manifestations and mimics

Consider, once again, a type of observable event, like the splintering of a macroscopic solid. This is the typical manifestation of fragility: the disposition to fracture or suddenly disintegrate when struck. Sturdiness or hardness is the opposite of fragility. It is a macroscopic solid's disposition to preserve its integrity when a certain degree of force is exerted on it.

However, as we have seen in the previous chapter, that a sturdy object is not disposed to break does not mean that it cannot break. It can break under the influence of some extremely strong force; or some force exerted from a specific direction to a specific point on its surface. Sturdy objects can have 'Achilles heels'. Yet, this does not make them fragile. Similarly, a fragile object can luckily fail to fracture even if a relatively strong force is applied on it, if it is exceptionally 'resistant' to that force at one specific point on its surface.

When a sturdy object breaks it 'mimics' fragility, and when a fragile object fails to break even if it was struck, it 'mimics' sturdiness.[1] In such cases, mimics are possible because sturdy objects *can* (are able to) splinter, while a fragile object *can* (is able to) resist splintering under certain uncharacteristic circumstances.

---

[1] As was mentioned earlier, the term 'mimic' is due to Johnston (1992). But the problem was first discussed in detail by Smith (1977).

Most objects have both the (generic) ability to break and the (generic) ability to not break even when a relatively strong force is exercised on them.

Sturdiness and fragility are incompatible *dispositions*: an object cannot be disposed to break and disposed not to beak at the same time; though it might be more or less disposed to break. However, having some generic ability to break and having some generic ability not to break *are* compatible. In fact, most solid macroscopic objects have both abilities. And it is precisely because they have both abilities that their dispositions, fragility and sturdiness can be measured and 'graded'.[2] The grades are determined by objects' different specific abilities to break when a certain amount of force is exerted on them in a certain manner (from certain direction, on a certain point of application, etc.).

But while almost every solid composite object can break and hence has the *unspecified* generic ability to fracture or disintegrate, they do not all have the same ability to break under the same forces in the same type of circumstances. We can distinguish those abilities through their connection to the counterfactual conditionals which their ascription entails. By using counterfactual conditionals, we specify more precisely the circumstances (in this case, the amount of force exerted, its direction, etc.) in which the object would break. The type of event which manifests the exertion of an ability is not sufficient for its specification.

However, the problem related to the phenomenon of mimics goes deeper than the issue of gradeability of certain modal properties does. Some landmines suddenly fracture or disintegrate when a relatively weak force is exerted on them. Thus, the conditional that the mine would suddenly splinter if it were hit or pressed with a certain of force is true. Nonetheless, we would be reluctant to ascribe the same disposition to a landmine and to a China vase. Mines are not fragile objects – in fact, they are sturdy – even if it is advisable to handle them with care. The splintering mine only 'mimics' fragility.

What this example shows is that we cannot distinguish the property of fragility and the property of being an explosive merely with reference to some generic type of event that would manifest them. Moreover, we cannot distinguish them even if we include their typical 'stimulus conditions' in their specification. It is interesting to observe that this problem is often disguised by the way we use verbs describing events in natural language.

I described the manifestation of both modal properties, somewhat artificially, as a (sudden) splintering of a solid object. I used this language exactly because it is neutral with regard to which modal property we may ascribe to the object.

---

[2] About gradable dispositions see Manley and Wasserman (2007, 2008).

However, this is not what we normally do. We would not call a landmine's splintering a *breaking*. And similarly, we would not call a fragile object's splintering an *explosion*.

What this linguistic practice clearly indicates is that, normally, we do not specify a disposition or an ability merely by some changes which an object (or its environment) would undergo whenever its ability or disposition is manifested. We naturally think of breaking as an event which manifests fragility. Sturdy objects are able to break while fragile objects are disposed to break. But neither of them is disposed or able to *explode* because they are *fragile*.

According to what I called the 'intentionalist view', modal properties can be identified simply with reference to those type of events at which a disposition or ability is 'directed'; that is to say, the events that would manifest them.[3] It seems to me that the appeal of the intentionalist view derives from the feature of our language that *the use of many (in fact, most) event-terms already presupposes the disposition or ability* that the events which they describe manifest. Fragility is the disposition (and/or ability) to break, solubility is the disposition (and ability) to dissolve and so on. However, the intentionalist approach is applicable only in those cases in which the use of an event-word entails that what happens is the manifestation of some already known disposition or ability. In such cases, knowing what happens presupposes knowing which modal property is manifested.

It is also important to note that many verbs' meaning is *ambiguous* precisely because they can be used to express the exercise of different kinds of abilities. Leaves can fly in the autumn wind, but they do not fly in the sense in which birds do. Jill can read a novel and she can also read the few Latin words which occur in it. But since she has never learned Latin, she does not understand them. Reading in the sense of understanding manifests a different ability than reading in the sense of being able to spell out some words. Some verbs have several meanings (or at least distinct shades of meaning) because we can use them to describe events which manifest different abilities. Thus, the way we identify an event is often essentially linked to which modal property is manifested or exercised.

John Mackie once described a situation which helped him to introduce his version of the conditional analysis of causation. But Mackie's example can also be used to illustrate how tightly the choice of an event-verb is connected to its

---

[3] Just as some mental states are identified by their 'intentional object' at which they are directed, as we have seen in Chapter 1. Importantly, again, this similarity does not mean that intentionalists about dispositions would consider every modal property mental. A possible consequence of this similarity is rather that intentionality is *not* the mark of the mental as Martin and Pfeifer (1986) seems to hold.

explanation and to the abilities that objects exercise when an event happens. In the example, a chestnut which is on a hot sheet of iron is struck by a hammer. Then the chestnut falls into pieces and its pieces scatter around (Mackie 1974, 29). We want to understand what exactly happened in this situation.

Perhaps the hammer exercised its ability to *break* a chestnut. Perhaps the hot iron exercised its ability to make the chestnut *explode*. In the first case, the chestnut broke; in the second one, it exploded. Perhaps the chestnut *both* broke and exploded (although it is unclear whether this form of 'overdetermination' makes much sense). Perhaps the breaking 'preempts' the explosion, perhaps the explosion 'preempts' the breaking. What matters for my current concern, however, is only that the question, 'Which ability has been exercised in the chestnut-case?' can be intelligibly raised; and that *the correctness of our description* of what happened (a chestnut's breaking or a chestnut's exploding) depends on which of its abilities has been exercised.

The chestnut case illustrates well why it is so easy to mimic abilities. We can describe many – although, as we shall see, not all – events as observable qualitative changes ('the sudden scattering of an object' let us say). But the occurrence of an event so described might involve the exercise of different abilities. Which ability has been exercised is not determined merely by which observable qualities change. By identifying an event as an exercise of an ability or as a manifestation of a disposition we can also explain what happens in the world.

Ryle once observed that we can explain the event that birds are flying south by saying that they migrate (Ryle 1949, 142). Similarly, we can explain an object's sudden splintering by its fragility, and we can explain its sudden splintering by its being an explosive.[4] For this reason, by re-describing an event, we can also provide an at least partial explanation of why the event happened. Whether the disintegration of the chestnut was a breaking-event or an explosion-event is determined by which ability has been exercised; or, from another perspective, which object's – the hammer's or the sheet of iron's – ability has been exercised.

Now the following seems to be true of Mackie's example: if, in the specific situation, the sheet of iron had been cold, the chestnut would have been *broken* (i.e. it would not have exploded); in contrast, if the hammer had not struck, the chestnut would have been *exploded* (i.e. it would not have broken). In my view

---

[4] Ellis and Lierse (1994) suggest that in order to specify (nomic) dispositions we need to suppose that there are natural kinds of processes. (On natural kinds of processes see further Ellis 1999, Chapter 2.) Whenever a disposition is manifested, a kind of natural process is going on. This means that the splintering of fragile objects and the splintering of an explosive must be two distinct kinds of natural processes. I certainly agree that the two types of splintering are distinct kinds of processes. But I wish to add that they are distinct types of processes precisely because they manifest distinct abilities.

– though not in Mackie's – the truth of these counterfactuals is grounded in the chestnut's abilities and what those abilities entail about the circumstances in which they would be exercised. Thus, what in fact happened is, *conceptually and not causally*, determined by which ability has been exercised.

Some events can of course be described without some tacit supposition about which ability they manifest. We can describe an event as a sudden splintering or a disintegration of a macroscopic solid. The splintering can involve the exercise of very different abilities and can manifest different dispositions. Some objects splinter because their splintering manifests fragility; while others do because their splintering manifests their ability to explode. It is for this reason that some of them splinter *because* they explode, while others splinter *because* they break.

In order to properly see the difference between the two kinds of cases, we have to rely on counterfactual conditionals which specify the circumstances in which they would splinter; circumstances which include reference to the absence of some factors that would mask their dispositions or their more generic abilities. Suppose we wrap the chestnut into a material which protects it when it is hit, but which is a good thermal conductor. Then, it is clear that in a situation, similar to Mackie's, the chestnut would explode rather than break.

This example suggests, again, that we cannot identify modal properties without linking them to some conditional. For it seems that we can often distinguish generic modal properties – abilities as well as dispositions – only by explicitly including reference to the absence of some relevant inhibitory factors in the antecedent of the conditional which specifies them. However, the phenomenon of mimics, just as the phenomenon of masks and finks, is not a mere technical difficulty about the conditional analysis of dispositions and abilities. It also presents us with some more interesting metaphysical problems, which concern the very notion of manifestation as well as the nature and identity of events.

As we have just seen, the classification of events often depends on which abilities they manifest. This important metaphysical connection can be easily missed if our investigation of modal properties is epistemically motivated. For it is certainly true that, in order to identify abilities or dispositions, we often need to observe how substances and objects change in certain (observable) circumstances. Given some minimal empiricist commitment, observations must play some role in the identification of modal properties. But this does not mean that abilities and dispositions can be specified merely with reference to observable episodes or changes.

When objects' abilities are exercised, some events happen or occur. There is a sense in which such events manifest objects' (substances', systems of objects') abilities or their dispositions. Which events can count as *manifestations* of modal properties is, however, a difficult matter. Manifestations are often understood as events which involve *qualitative changes*.[5] Qualitative changes have *originally* been understood as events which can be observed. The language in which such events are described is the language of *appearances*. But as we shall see in a moment, it is far from obvious that the results of the exercise of every modal property could or should be understood in terms of an (in-principle) observable qualitative change.

In the sequel, I shall argue that we need to distinguish two distinct ways in which we can classify events. Events can be classified as certain types of *observable episodes of changes*. But events can also be classified with reference to which modal properties they manifest. This means that we cannot classify events independent of how we specify abilities. Events and changes are not ontologically prior to the modal properties which they manifest. This means that events are what they are because, essentially, they are the results of the exercise of a specific sort of ability.

Moreover, not every disposition or ability is manifested by the changes of the object which bears them. Fragility is the object's ability (or rather: disposition) to break in certain circumstances; solubility is the object's ability (or again: disposition) to dissolve in certain types of solvents. But in Mackie's example the hot sheet of iron does not – or at least need not – *change* in order to exercise its ability to make a chestnut explode. In fact, abilities that are manifested by the changes of its bearers are rather special. We often call them liabilities; though liabilities, on the account I argue for, are the same sort of modal properties as abilities. They ground more or less abstract possibilities.

Aspirin has the ability to relieve me from my headache. But this ability is not manifested by a change in the aspirin, even if its exercise does involve the change of the state of the aspirin. A piece of magnetized object's ability is exercised when it changes the position of the needle of a nearby compass. But the magnetized object need not undergo any change in order to exercise that ability. Most often, though not always as we shall see, when abilities are exercised some change takes place in the world. But the exercise of an ability need not be manifested by a change *in* the object that possesses it.

---

[5] See especially Lombard (1986).

Events, states and abilities

So, what happens when an ability is exercised? I suggest that what happens is that a more or less abstract possibility is realized by a concrete event: something that happens at a certain time in a given region of space. And, as I argued in Chapters 2 and 5, whenever a disposition is manifested, some abilities are exercised as well.

The account of abilities and events I am arguing for is reminiscent of Aristotle's account of motion and, more generally, change (*kinesis*) in *Physics*. According to Aristotle's definition, change is 'the fulfilment of what is potential as potential' (*Physics* 3.1, 201b5). The precise meaning of this definition is a matter of scholarly debate.[6] But one way to interpret it seems to be this: events that happen in the world are what they are in virtue of being realizations of possibilities; which are, further, grounded in objects' dynamical-modal properties (potentialities, as the standard translation of Aristotle says). That is to say, which *sort of* change happens at a particular time in the world depends on which abilities are exercised.

How can we identify the relevant possibilities and objects' abilities that ground them? I have already mentioned some arguments which speak against intentionalism as an account of the identification of modal properties. But the strongest among them is related to the phenomenon of mimics. For what the possibility of mimicking ultimately shows is that the specification of events that manifest an ability is not independent of the specification of the abilities themselves.

As we have seen earlier, there are cases in which the 'stimulus' conditions of different abilities are the same (kind of) events; and the manifestation events described *merely as a qualitative change* are also the same (kind of events). A landmine or a (cocked) hand grenade is not fragile, even if they would suddenly disintegrate as a response to being hit. But they would not break; they would explode. Breaking is not the same type of event as exploding. Fragile objects have the disposition, and hence the ability, to break; explosives have the disposition, and hence the ability, to explode. Certain instances of disintegration will be the exercises of the ability to break, while others will be the exercises of the ability to explode.

---

[6] In the early modern period, Aristotle's definition was often parodied as an example of the 'eminent trifling in the Schools'; see Locke (1689 Book III, iv, 7–8). In contrast, Leibniz clearly understood the meaning and significance of this definition; see the relevant section in Leibniz (1765) (Leibniz's response to Locke). For scholarly interpretations about Aristotle's account of motion and change see, among others, Kosman (1969) and Charles (1984), Chapter 1.

This shows, more generally, that what happens in the world depends on which abilities are exercised. A change is, as Aristotle said, the 'actualization of a potential'. I would say, more generally, that an event is the realization of an abstract possibility by some concrete object(s) at one spatiotemporal region. This means that an event is what it is because it manifests an object's, substance's or some system's ability.

It is of course true that in many cases, when abilities are exercised, qualitative changes might be observed as well. Nonetheless, there is a sense in which every qualitative change is only a 'mimic' since such changes themselves can reveal *the exercise of many different underlying abilities*. Often, when we describe an event as a qualitative change, we still want to understand what is *really* going on.

In fact, purely qualitative changes which manifest dispositions in the original, empiricist-phenomenalist sense of manifestation can specify only one sort of ability: physical objects' ability to elicit certain observational responses in us. Suppose that disintegration is understood as a merely qualitative change which can be the (intentional) object of our observations. Yet, fissions, breakings and explosions – to mention just a few types of disintegration – are going to be different kinds of events or processes depending on which abilities are exercised. Merely qualitatively described changes, like observable changes of motion and shape, are not sufficient to identify an ability in most cases.

What explains the difference between breaking and exploding is that sudden disintegration as a qualitative change can be the result of the exercise of different substances' abilities. An object which contains some explosive material, like a hand grenade, would expand with a high speed if the object were hit and/or if pressure were exercised on it. A fragile object would fall into two or more pieces if certain kind of force were exerted on it from certain direction. This is only a rough approximation of the proper conditionals that specify these abilities. Nonetheless, the point is that we can distinguish events only by specifying which abilities objects exercise, and we can specify the abilities with the help of linking them to more complex and precise conditionals.

Moreover, qualitative changes are not even necessary for exercising abilities. As far as inanimate nature – that is, matter and material changes – is concerned, the most obvious and important ways of exercising abilities without *qualitative* changes in the objects are motion and rest. If motion is understood as change of relative position, motion must in principle be possible without qualitative change. But even if every motion did involve some qualitative change – for instance, if the length of an object did change when it moves relative to another – this change would not constitute the exercise of an ability; at most, it could be a *further consequence* of its exercise. Locomotion is a (perhaps *the*) fundamental form of physical changes. Hence, the exercise of many physical abilities need not result in *qualitative* change.[7]

An accelerating body changes its velocity as well as its ability to do work. Its velocity might not be a modal property, but its ability to do work certainly is. But neither change of the object (physical body) is 'qualitative'. Nonetheless, we are able to understand them perfectly well. It would be very hard to distinguish the distance of electrons from the nucleus in an atom on the basis of the arrangement of their 'qualitative properties'. Yet, we understand what those particles are and what it means for them to be a certain distance from the nucleus. (Well, we might not understand it perfectly well, but that has nothing to do with the relevant changes not being 'qualitative'.)

Animated nature is full of examples when we ascribe abilities the exercise of which are manifested by 'qualitative non-changes'. Consider a healthy person with a well-functioning immune system. The immune system's ability will be exercised when a person is *not* experiencing any particular change; that is, she fails to develop any (observable or experienced) symptoms of some disease even when she is exposed to various types of viruses or bacteria.

The ability of an organism having a well-functioning immune system is exercised when the organism is protected from various kinds of dysfunctional changes. But we cannot specify the organisms' ability by saying that they do not display symptoms of some disease. Minimally, we need to add that they would not display that symptom even if they were exposed to some kind of bacterial or viral infection; otherwise, we should ascribe the same ability to people with a bad immune system who live in a luckily safe environment as we ascribe to the healthy ones. This means, again, that we cannot specify abilities without connecting them to the truth of certain counterfactual conditionals.

An organism with a well-functioning immune system is a rather complex entity, and one might object that the specific biochemical processes which subserve the working of an immune system must involve some 'qualitative' change. However, this can hardly prove that the exercise of an organism's ability must involve qualitative change too, unless the sub-serving processes themselves can be identified without reference to their ability to contribute to the preservation of health. But we cannot understand what the relevant abilities of the biochemical

---

[7] Perhaps if physical space is understood as a substance (or as a 'container of the material bodies') then we can conceive the rearrangement of objects' relative position in space as a sort of qualitative change of the spatial substance's structure. But even then, it is hard to conceive the spatial arrangement of particles or bodies as 'qualities' of the space.

constituents of a living organism are without an essential reference to their contribution to the 'non-changes' of an organism as a whole.

Rom Harré once said of powers that they 'may be in opposition in such a way that though much is going on, nothing actually happens. The powers of greed are not abrogated in the case of Balaam's Ass, they are exercised, but still nothing happens, and the explanation of the fact that nothing happens is given in terms of the exercise of power' (Harré 1970, 93–4). Harré's claim that 'much is going on' but 'nothing happens' sounds paradoxical, but it makes perfect sense if we distinguish qualitative changes and exercises of abilities: abilities can be exercised (much is going on) even when 'nothing happens' in the sense that no qualitative change occurs.

Some 'non-changes' can manifest abilities because, following Harré's wording, the exercise of certain abilities can be 'in opposition'. Some abilities are 'in opposition' only in some peculiar circumstances, but there are other kinds of abilities that are essentially preventive: they are exercised when they *inhibit the manifestation* of some other abilities. Consider, again, the preservation of someone's health by the operation of her immune system. Although nothing 'happens' to the person in the sense that she is not changing in respect of her health, this 'non-change', in certain circumstances, counts as a manifestation of the ability (or abilities) of her immune system.

Standing motionless like a pantomime statue does for ten minutes is obviously the exercise of an artist's ability, even if no change is involved. So is having a sleep; or holding back our breath. Apples are hanging peacefully on a tree. Nothing happens. But that they are not falling on the ground is the manifestation of the ability of their stalk: it is exercised by 'opposing' the Earth' ability (or in this case, we may say, power) to exercise gravitational attraction on objects on its surface. In all such cases, 'nothing happens' in the sense of no change is occurring; but 'a lot is going on' in the sense of many abilities being exercised.

More generally, the exercise of preventive abilities and abilities to resist some change is manifested by some 'qualitative non-change'. Perhaps some non-changes – although certainly fewer than we would initially think – do *not* manifest the presence of any ability. But most of them manifest the exercise of *many* abilities. It is a mistake to hold that 'non-changes' or 'standing states' are the 'normal' states of the world so that the exercise of modal properties should explain only changes. In fact, to understand states of non-changes we need as much modal properties in ontology as we do for understanding changes.

Thus, not only qualitative changes but also qualitative states understood as temporal intervals of non-changes (like being healthy or having a cubic shape) involve the exercise of many abilities. The exercise of abilities often supports states of equilibrium that are manifested by qualitative non-changes.[8] In fact, every qualitatively non-changing macroscopic solid is the manifestation of the exercise of the abilities of (the milliards of) its constituents.

Since many abilities are exercised by some 'qualitative non-changes', it seems that the exercise of an ability is not always an event. But this follows only if we insist that events must involve qualitative changes. We can also argue inversely that, since many abilities are exercised when no such change occurs, some events are not qualitative changes. Perhaps an event is *essentially* an exercise of an ability, which can be manifested either by some qualitative change or by some non-change in specific circumstances.

If so, then 'qualitative non-changes' are events which manifest the joint exercise of many abilities. We can understand preserving colour or shape, resistance to certain forces, staying alive or healthy, and so on as non-changes which are the *results* of exercising many abilities.[9] Yet, these non-changes are the results of such events which we can identify with the help of the same sort of conditionals that we employ when we identify the abilities themselves.

For instance, we can identify an ability by the conditional that, roughly, if an object o were exposed to certain kind of force exercised on it by another, it would resist a certain kind of change Φ until a certain time t. The result of the exercise of an object's ability can be described as a state. In cases of resistance, the state might be the same as it was before the exposition of the relevant force. But that which manifests the object's specific modal property must be an event: in this case the event of resisting a certain kind of change.

Suppose I try to scratch first a glass, and then a diamond. The glass changes (becomes scratched), and that is certainly an event. How could we deny that the diamond's resisting to being scratched in similar circumstances is also an event? Do we really want to say that when I exercise the scratching-type force on the diamond, nothing is going on with it? This I find completely implausible. Something must be going on; something which can make it the case that the

---

[8] Given that the interactions of abilities can result in non-changes, it is logically possible that there is a universe without *any* qualitative change, where nonetheless many abilities are exercised. This would, in a sense, be a universe in 'perfect equilibrium'. A universe in this state might explain how a 'frozen universe' of the kind that Shoemaker describes in his famous thought-experiment about time without change could be possible (Shoemaker 1969). When the universe is in such a state, 'nothing happens' in the sense that no observable, qualitative change occurs. Of course, it is not true of this universe that there is 'nothing is going on' in it; otherwise, it would be impossible to explain how the frozen world can start changing qualitatively after a certain time.

[9] As Aristotle says, one kind of potentiality is the ability of being acted upon; another kind is 'insusceptibility to change for the worse and to destruction', *Metaphysics*, Book IX, 1, 1046a, 10–15.

diamond remains unscratched under these circumstances. To think otherwise would mean that there are (physical) actions *without* reaction. The non-change in these specific circumstances/conditions is an event, and it is an event which manifests the diamond's ability, and which is identified *as* the exercise of that ability. The non-change reveals something which is going on. It could not be the mere fact that the diamond was unscratched at time *t*, which would be true even if it had not actively participated in any physical interaction.

Finally, it is also important to note that there are many sorts of changes which may or may not involve qualitative change, but which are not changes *merely* in virtue of their qualitative features. The ability of a certain amount of arsenic to kill a person who digests it (without taking antidotes in due time) will result in a death. But whatever death – or birth – is, they are not *merely* qualitative changes. In fact, it is interesting to note that many of Aristotle's examples of the actualization of potentialities are generations of individual substances (instances of kind of objects or organisms).[10] This kind of change is obviously different from the merely qualitative changes of an already existing object or organism.

Thus, we cannot identify abilities merely by linking them to some sort of observable qualitative change. We need to use counterfactual conditionals to distinguish different types of resistance to change, preservations of some quality and so on. In such cases, many abilities are exercised on the same occasions. However, we have already seen in the previous chapter that this is not exceptional. Qualitative changes too almost always occur as results of many abilities exercised together. A sugar cube dissolves in water: its ability to dissolve is exercised together with the water's ability to dissolve sugar. A fragile object breaks when it is hit by a hard one: its ability to break is exercised only when another object's hardness is. When a sturdy object resists breaking as a response to being hit by a hard one, its ability is exercised no less than a fragile object's when it breaks.

Manifestations and effects

Being 'manifest' is the opposite of being 'latent'. What is 'latent' is hidden from our senses. What is manifest is open to our senses. According to the traditional phenomenalist accounts of modal properties, objects' dispositions are manifested when they become, in the literal sense, sensibly accessible to us.

---

[10] For instance, a house or a man. See *Metaphysics* Book IX, 7 and 8.

For most of the early-twentieth-century empiricists, the manifestation of a disposition was not an *effect*. These philosophers were the followers of Hume not only in his phenomenalism but also in his scepticism about causation.[11] For many logical empiricists, causation was either a conceptual remnant of some uncritical (and unscientific) language; or a mere by-product of the exact, scientific-nomological understanding of the world. In either way, they thought that causal concepts cannot play any *fundamental* role in philosophy (or in philosophy of science, especially).

Consequently, logical empiricists did not attempt to analyse modal concepts like dispositions with reference to causation. They were aware that 'disposition terms' are frequently applied in the language of science, but they did not interpret such terms as causal concepts. And although Ryle rejected phenomenalism, he still interpreted dispositional explanations as *noncausal*. In fact, the analysis of disposition terms played such an important role in his account of the mind because he thought that interpreting mental discourse in terms of behavioural dispositions can avoid the flaws of the traditional Cartesian causal-interactionist metaphysics.

Historically, the rejection of noncausal theories of dispositions was a response to Ryle's account of the mind. David Armstrong's materialist metaphysics agreed with Ryle – and with Wittgenstein, to the extent that this was what Wittgenstein really meant, which is always a controversial issue – that mental properties are dispositional. But he also insisted that the Cartesians were right to hold that the mind causally interacts with the body. In Armstrong's view, these 'Cartesian intuitions' are consistent with a dispositionalist account of mental properties, if we interpret the behavioural manifestations of mental dispositions as *effects* of inner causes.[12]

However, it was a crucial assumption of Armstrong's theory that the inner cause of behaviour is not the disposition itself. When we ascribe dispositions to agents, we opaquely refer to some kind of non-dispositional, 'categorical state'; specifically in the case of human action, to some kind of neural state. It is these 'categorical states' that are the causes of behaviour which manifests a disposition. Dispositions have effects only in virtue of 'being identified with' or 'being realized by' some categorical state or structure.

---

[11] The most famous expression of this attitude is, of course, Russell's injunction that 'the law of causality' is 'a relic of a bygone age, surviving, like the monarchy, only because it is erroneously supposed to do no harm'; Russell (1976, 173). Another, very influential, strategy of the early-twentieth-century empiricists was to revise the traditional ('everyday') concept of causation and reinterpret causality as a special form of nomic connection. See for instance Schlick (1932).

[12] About manifestations as effects see also McKitrick (2010).

I shall call the accounts according to which the manifestation of a disposition or the exercise of an ability is the effect of some underlying causal process *causalist* theories.[13] The causalist theories of dispositions should not be confused with the dispositionalist accounts of causation. Some philosophers argue that the very concept of causation should be understood as an exercise of dispositions or of 'causal powers'.[14] Those whom I call causalists reject the dispositionalist accounts of causation.

Causalism, as I use the term here, refers to the idea that manifestations are effects. But this does not mean that causalists would think, like the dispositionalists do, that manifestations are caused by an object's modal properties (typically, its dispositions or powers). Rather, causalists believe that manifestations are caused by a 'triggering event' (like the striking of a piece of glass) and the 'categorical basis' of the modal property (the glass's crystalline structure) which is assumed to be nonmodal, and which jointly causes the manifestation (the breaking of the glass).

In fact, some causalists even argue that dispositions are 'causally inert'.[15] On their view, the manifestations of dispositions are effects of some causes which must be distinct from an object's disposition. Manifestations cannot be the effects of modal properties, since effects *described as* manifestations are conceptually connected to the dispositions which they manifest. Hence, if we grant the Humean assumption that effects cannot be conceptually connected to their causes, dispositions cannot be causes at all.[16]

In this section and the next, I shall argue against a causalist interpretation of modal properties. The exercise of an ability need not be manifested by an effect. While exercising abilities often involves the operation of some causal mechanism, *manifestations as the results* of exercising abilities ought not to be understood as effects. The manifestation of fragility is sudden fracturing, but sudden fracturing is not caused by 'the categorical properties and structures' of the object; neither, in fact, by its fragility. It is caused by the fragile object's being struck. The manifestation of a substance's being toxic is someone's getting

---

[13] Apart from Armstrong (1968, 1973), the most important representatives of such accounts are Mackie (1973, 1977), Prior et al. (1982), and as we have seen, Lewis (1999/1997).

[14] Among others, Harré and Madden (1975), Molnar (2003), Martin (2008) and Mumford and Anjum (2011).

[15] Especially Mackie (1977) and Prior et al. (1982).

[16] About the so-called conceptual connection argument which is meant to support the idea that dispositions are 'causally inert', see especially Mackie (1973, 136–7; 1977). If the argument were right, it would of course undermine Armstrong's original idea that the causal reinterpretation of dispositions can explain the possibility of mental *causation*. But the argument is not sound; it confuses the question of how events are related to corresponding properties with how they might be related to other events (see Mellor 1974, 179–80).

poisoned, but the cause of getting poisoned is either the toxic substance or the event that the substance has been ingested.

This does not mean that modal properties are 'causally inert'. But it is one thing to deny that a property is 'causally inert'; and it is quite another to assert that *properties* are causes; or that modal properties are the causes of their own manifestation or exercise. Causes, at least 'efficient causes' of effects, are either objects or chemical substances; or events in which such objects and substances are involved. Substances might be causes, or they might be involved in some sort of causal interaction, only in virtue of having some properties; but neither abilities nor 'categorical properties or structures' (if we assume, just for the sake of argument, that there are such things) *are* causes in the relevant sense.

When a certain object' surface is red in a region r at time t, its being red there and then instantiates the property of redness. But redness does not thereby cause the object's being red in r at t. Analogously, an object's sudden disintegration at t can be an exercise of its ability to break or its ability to explode, but those abilities do not cause its disintegration. Rather, they are exercised when the relevant event happens. The relationship between a property and its instance or a modal property and its exercise is conceptual *and* metaphysical, not causal.

Be this as it may, my interest here is not causation or the general role of properties in causation. Rather, I want to call attention to a crucial shift in the meaning of what philosophers called the 'manifestation' of a disposition. For the phenomenalists, manifestations are indeed *appearances*; private episodes of observations. For Ryle, 'manifestations' are a kind of publicly observable behaviour of some objects rather than private episodes in the mind of observers. But for the causalists, manifestations are *effects* of some causes *irrespective* of whether or not such effects are directly accessible to the senses. The disposition of a conductor is 'manifested' by its conducting electricity well. But that electricity is conducted well is not directly accessible to the senses as the dissolution of a sugar cube in my coffee might be.

As I have mentioned in Chapter 4, early discussions of dispositions were driven by semantic and epistemic considerations. The logical empiricists' aim was to provide an analysis of the meaning of disposition *terms*. We cannot simply see (or hear, smell, etc.) the solubility of a sugar cube as we can its (spatial) unity, its shape or its colour; nonetheless, we can correctly classify objects in science as being water-soluble or as not being water-soluble. What justifies such classifications is that objects' dispositions can (literally) *become manifest* by observing how they behave in certain circumstances. Thus, it was an epistemic challenge which motivated the semantic agenda: the attempt to provide an analysis of the meaning of disposition terms.

For a phenomenalist, a modal term is meaningful only if it can be logically linked to sentences reporting what is observable. But when the causalist program gradually took over the phenomenalist one, the notion of manifestation was implicitly reinterpreted as a *response* or an *effect* to some stimulus, rather than something that is, by definition, observable. Some responses or effects can, of course, be directly observed. But some modal properties can be exercised without any observable changes as a result, since the relevant 'causal responses' may or may not be directly observable. Paradoxically, manifestations understood as responses to stimulus might not be literally *manifest*. Some batteries are rechargeable. But when they are actually recharged, we cannot *see* the effect. Effects can occur even when they are not directly observable.

Nonetheless, causalist accounts of dispositions inherited one important aspect of the former 'phenomenalist' accounts. This is the very idea that the content of a modal property (typically, of a disposition) depends, at least partly, on the species of events, which are their manifestations, rather than the other way around. Thus, although for different reasons, both phenomenalists and causalists thought that the nature of the events which manifest a modal property is metaphysically independent of which modal property is manifested.

But as I argued in the previous section, the phenomenon of mimics and of different forms of non-changes clearly proves that both views are mistaken. For what happens – that is, which kind of events occur – often depends on which modal property is exercised. The correct identification of what happens often goes hand in hand with the correct identification of the ability that is exercised.

Nonetheless, one might still insist that the events that manifest modal properties must be effects. For it might be argued that events that are not (or not only) qualitative changes can be identified by their *causal origin*. What sort of event happens then does not metaphysically depend on which ability is exercised, but rather on what sort of thing caused it. In the next section, I shall argue against this view.

Causal consequences versus abilities' results

To begin, let me return to a rarely mentioned aspect of David Lewis's 'reformed conditional analysis'. As we have seen in Chapter 4, it is a crucial feature of Lewis's analysis that the consequent of the counterfactual conditional which is logically linked to the ascription of a disposition is not the occurrence of an event. The consequent of the conditional is that a 'response' would be jointly *caused* by the occurrence of the 'stimulus' and the object's having the property that is the disposition's 'intrinsic base'. This is a major divergence from earlier versions of the conditional analysis, according to which an object o has the disposition to φ if and only if o would φ if it were in circumstances C.

Lewis's is a causalist interpretation of dispositions, but it seems to involve a reinterpretation of the very idea of manifestation as well. For in Lewis's analysis, it is not simply the occurrence of a type of event which manifests a disposition. What manifests it is the *causing of an event* (or 'response') by the 'stimulus' and the object's having the relevant intrinsic base. Hence, manifestations are not events at all, but *the causings* of events. On the previous causal interpretations, in contrast, manifestations are the events that are *the effects* of some inner causes. It is worth dwelling on this difference because they are closely related to the issue whether events are conceptually/metaphysically dependent on the modal properties which they realize.

The idea that a manifestation is a causing of an event rather than an event which is the effect of some inner causes is reminiscent of a debate over how to understand the intentionality of human actions. According to the so-called causal theories of actions, the intentionality of behaviour is explained by its causal origin; where the cause of an action is either the agent as a substance or some of the agent's mental state(s) like her beliefs, desires and intentions. But while, on some such accounts, the movements of the agent's body are her actions if they *have been caused* by the agent or by some of her mental states, according to others, bodily movements are events that are not actions, but the *results* of an agent's actions. Actions themselves are the *causings* of the bodily movements.[17]

On this latter view, my arm's *rising* is not an action; although it may or may not be a result of my action. *Raising* my arm, in contrast, is an action because raising an arm is my causing it to rise. Applying this distinction in the context of abilities, we can say that the movement of my arm may or may not be an exercise of my ability to raise it (intentionally). It is my causing the arm to rise in a specific way – for instance, by wanting or intending that it goes up – which manifests my ability to move my arm.[18]

---

[17] Davidson (1980/1971) argues for the first view; as a response, see Alvarez and Hyman (1998), which argues for the second.

[18] Notoriously, causal theories of action must face the problem of 'deviant causal chains'. The problem arises because it is possible that the agent's proximate intention to φ can cause her φ-ing even if her action was not intentional. To mention a standard example, a novice on the stage might intend to tremble her hand in a particular moment, but her having that intention makes her so nervous that it

What kind of abilities might be manifested by an agents' bodily movement is a contentious issue in philosophy of action, which I cannot and need not pursue here. But it is important to note that in the two kinds of cases – that is, in the case when the bodily movement is interpreted as the agent's intentional action and when it is interpreted as the action's result – *different abilities are exercised*, since the agent can have the *generic* ability (or capacity) to perform a bodily moment of type B, irrespective of whether it was her intentional action. If performing an action intentionally is the exercise of an ability, which is different from the ability exercised in cases of nonintentional bodily movements, then we cannot classify a bodily movement *simpliciter*, as if it were the same kind of event in both cases. What we should say is rather that *different kinds of events* occurred because different abilities have been exercised.

This shows, once again, that which kind of events occur cannot always be correctly identified without specifying which abilities are exercised even if the exercise of distinct abilities is *manifested* by the same 'qualitative change' (bodily movement). Since both intentional actions and unintentional movements can manifest *some* ability, the movement of the arm can be considered as the manifestation of either of them, but not both. Thus, we cannot properly distinguish certain abilities merely with reference to qualitative changes that occur when they are exercised.

Nonetheless, one might insist further that even if qualitative changes are not sufficient to identify what happens (or 'what is going on'), events' causal history might be. Following Lewis, one might claim that a disposition is 'manifested' – or an ability is exercised – when the 'stimulus' and the object's having of the intrinsic base property would jointly cause the occurrence of a type of event. If the 'response-event' were caused by a different stimulus-event or a different intrinsic causal base, then a different ability would be exercised; or perhaps no ability would be exercised at all. For what manifests the ability is the causing of an event by the *appropriate* stimulus and the object's having the *appropriate* intrinsic, nonmodal base-property.

Alternatively, we can deny that two events that manifest distinct abilities are of the same type on the ground that what events are is determined by their causal history.[19] Originally, the causal theory of events concerned singular events

---

causes her hands trembling unintentionally. In such and similar cases, the agent's proximal intention to perform an action causes her unintentional behaviour. Such cases may, in fact, undermine the whole idea that that relation between intentions and actions should be causally understood, and support an ability-based, noncausal understanding of intentional actions. This problem is, however, beyond the scope of the present study.

[19] See, for instance, Davidson (1980/1969) and van Inwagen (1978).

or events as particular occurrences. Nonetheless, it is arguable that *types* of causal histories might determine which *types* of events are instantiated. Getting sunburned may describe a particular event, but it can also characterize a type of event: the event of altering skin's colour as a causal consequence of being exposed to sunshine.

There is certainly some analogy between the ideas that types of events are identified by the types of their causes and that they are individuated by the kind of ability which they manifest. Nonetheless, the two views are different. First of all, as I have already mentioned, I do not think that modal properties themselves cause the events which manifest them, or which are the results of their exercise. Not because such properties of objects are conceptually linked to the events which manifests them – for conceptual and causal connections are, *pace* Hume and the Humeans, often compatible – but because abilities and dispositions are properties and I do not think that it makes sense to consider modal properties as causes of their own exercise.[20]

Moreover, although most abilities are exercised only if certain conditions obtain, it is a further question whether we should regard all these conditions as 'causal' in the sense that they 'stimulate' (that is to say, cause) the occurrence of a 'response'. A metronome is able to produce a sound at a regular interval. Thus, it is true of it that it would click precisely at some $t' \geq t$ if it clicked at t, no matter which specific time t is. But that it clicks at t is not a cause of its clicking at t'. One may or may not want to interpret the underlying physical mechanism of the metronome causally, but the condition which helps to specify this ability is surely not to be understood as a 'stimulus' for a causal process.

A good car has the ability to accelerate at the same rate on an elevating road as it does on the flat road. That ability is manifested only in those conditions in which a road is elevating. But an elevating road does not seem to be a cause of the car's acceleration (in the upward direction) in any sense. Some sort of causal process is, of course, involved in the exercise of the car's ability. But the salient condition of the ability's manifestation is not a 'stimulus', and it in no sense causes the event that manifests the object's ability.[21]

---

[20] Lewis seems to be assuming that 'x's having property B' is an event, which can be a cause. I must confess it is hard for me to make clear sense of this idea. In other places, Lewis takes events to be properties (Lewis 1986b). But how is this compatible with the view here that an event is 'having a property' and that having a property is a cause?

[21] Importantly, not every nomic ability is necessarily causal. The changes of an ideally closed system can manifest the system's ability to evolve in a certain way rather than another. The exercise of this ability is *not* conditionless. The conditions on which its exercise depends are the system's intrinsic properties; but these are conditions, nonetheless. For the system has the ability to turn into state S_2 by some time t', if it were in state S_1 at t, and it can have that ability even if S_1 never actually obtains.

Second, many abilities are manifested by *not being exposed* to certain causal influences rather than being caused to behave in a certain way. As I have already mentioned, the exercise of abilities can be manifested by the resistance to certain changes. One characteristic feature of solid bodies is their hardness. Hardness is the ability to resist certain kind of change, like being scratched, indented or broken. This ability is manifested when exerting a certain force on a solid object *fails to cause* certain changes in it.

There are many important physical, chemical and biological abilities that are manifested by a substance or organism's resistance to certain kind of changes. Gold has the ability to resist dissolution in (most kinds of) acids. Steel has the ability to *resist* rusting *longer than* iron. Resisting to some kind of force, or resisting another object's causing some changes, involves the exercise of an ability. So does the resistance of a person's body to the potential causal influence of certain bacteria and viruses.

Resistance to some kind of causal interaction plays an especially important role in *grading* some dispositions. Minerals' hardness, for instance, is measured by a scale identifying which kind of mineral can be scratched, or can resist to being scratched, by another kind of mineral. Resistance to being scratched by another kind of substance is the manifestation of a certain specific degree of hardness. Diamonds cannot be scratched by any other minerals; hence, they are the 'top (or the bottom) of the list'. They have the ability to scratch all other minerals, but they resist to be scratched by any other.

A knife made of stainless steel can have the ability to resist rusting and the ability to resist blunting. The ability to resist rusting and the ability to resist blunting are distinct abilities, even if the manifestation of those abilities does not involve the occurrence of any causal process. These abilities are exercised precisely when some type of causal process fails to occur. Since no causal process occurs when such abilities are exercised, the abilities cannot be distinguished with reference to their causal origin. They can be distinguished only with reference to the circumstances in which the absence of some change manifests an object's ability or disposition.

Thus, the identity of some (type of) events ontologically depends on which modal properties are realized by a concrete substance or object at a spatiotemporal location, and not on the event's causal origin. In some cases, abilities are manifested by the object's resisting certain forms of causal

---

But this does not entail that being in state S_1 (self-)causes the system to be in state S_2. All that needs to be assumed is that there is a nomic-functional dependence between the states. This does not entail that the former causes the latter.

influences. Resisting a sort of causal influence is a concrete event: it is something that happens in a certain space and at a certain time (or temporal interval), with a certain object (or substance, or system of objects).[22] But it is not the effect of a causal process. And further, even when the exercise of an ability does involve some causal process, an event is not identified *as* the effect of that process, but as the (noncausal) result of the exercise of a given ability. Metaphysically speaking, it is modal properties, and not some (observed or observable) qualitative change or some types of causes, which determine which kind of events occur and when in the world.

Counting events and abilities

So far, I argued that events can be classified both as qualitative changes, which may be directly observable, *and* as results of the exercise of some ability (or abilities). Events as qualitative changes and events as exercises of abilities are metaphysically different, but they are not entirely distinct. Otherwise, it would be a mystery how we can learn about objects' abilities and hence about 'what is going on' when something 'happens'.

In order to clarify the connection between the two types of events we need to raise a question with which Aristotle was already much preoccupied. The question is: how many events happen when objects (or samples of substances) interact?[23] A sugar cube has the ability to dissolve in hot water. My hot tea has the ability to dissolve sugar. How many events happen when I put a sugar cube in my tea? The initially plausible response might seem to be that only a single event happens: the dissolution of the sugar cube in my tea. Martin, for instance, claims that 'the reciprocal dispositional *partnering* and their mutually *manifesting* are *identical*' (Martin 2008, 46); and this suggests that, on Martin's view, in such cases the two dispositions (or abilities) are manifested by a single event.

But this answer seems so plausible only if we interpret events as qualitative changes. For if events are essentially exercises of abilities or the results thereof,

---

[22] The idea of events as manifestations of objects' abilities does seem to preclude the possibility of 'objectless' events. But since the possibility of objectless (or substanceless) events are incompatible with many other analyses of events as well, and since it is questionable anyway whether there is any example of objectless or substanceless events, this does not seem to be a particularly contentious assumption. For there do not seem to be events which do not happen to anything – be they chunks of matter, material particulars composed of them, persons or even perhaps immaterial substances, if there are such things.

[23] See especially Aristotle, *Physics*, Book III, 3 202b8ff. For an interesting discussion of Aristotle's notion of change, see Charles (1984), Chapter 1.

then at least two events must be going on: the sugar cube is being dissolved in water and the water is dissolving the sugar cube. If events are classified with reference to which abilities are exercised, then these two events cannot be the same. For the sugar cube's and water's abilities are different. Certainly, they are 'disposition-partners' in the sense discussed in the previous chapter. We need both water and sugar to exercise their respective abilities. But that does not make the very exercise of those abilities the same.

Further, consider preventive or inhibitory abilities. Objects or substances with distinct abilities can interact so that one of them is exercised when it prevents or inhibits the manifestation of another ability. In contrast, some objects or substances, like catalysers, have the ability to interact with others so that they 'assist' the exercise of their abilities. Alternatively, a manifestation might require 'interactions' of different abilities in the same object. The ability of a magnet to attract objects made of iron and the ability of the earth to attract an object with a certain mass might both contribute to an object's movement with a certain velocity (or with a certain rate of acceleration) in a certain direction.

To save the idea that when abilities are jointly exercised only one event happens, it has been suggested that abilities are not manifested by certain events, but by their *contributions* to the occurrences of certain events considered as effects. According to this view, abilities are similar to physical forces. Different forces can produce a joint effect. The net force exerted on an object can be calculated by certain rules like the parallelogram rule for vector addition. Each individual force has a mathematically calculable 'contribution' to the occurrence of the net effect.[24]

What forces are, and how similar they might be to abilities, is a contentious issue in the philosophy of physics. The issue is contentious because dispositions and abilities are modal *properties*, while forces need not be interpreted as properties at all. Perhaps forces are better understood as *transmitters* of some physical quantities.[25]

This does not mean that forces might not be associated with objects' or particles' abilities. For instance, in Newtonian mechanics, forces can change massive objects' momentum. Objects that exert different forces (e.g. gravitational and electromagnetic forces) can (are able to) produce an effect that neither

---

[24] The idea that manifestations are contributions to effects originates in Molnar (2003), 194–8. For a detailed development of this idea see Mumford and Anjum (2011), Chapter 2.
[25] According to contemporary physics, elementary forces are transmitted by force-carriers called bosons.

of them would produce individually. On the model of forces, we might try to understand the exercise of an ability as a sort of contribution to a joint effect.

However, it is not easy to understand how the exercise of an ability can in general be manifested by its 'contribution' to some effect. In many cases, abilities are not quantifiable properties. Cyanide would cause death if it were inhaled. But what could it mean to say that it 'contributes' to a person's death? The idea might be that cyanide can 'contribute' to someone's death only if it interacts with some specific kind of living organism. However, it is one thing to admit that most abilities can be exercised only if objects with other 'partner' abilities are present – this, as we have seen, is widely accepted among philosophers ever since Aristotle (and even reductionist about modal properties grant it); it is another to claim that *the exercise* of an ability *is* a contribution.

Contributions are not parts of events; neither are they some aspect of processes. Parts of processes are phases, which are other, shorter processes. But events are the *results* of the exercise of an ability, not parts of the process, if any, which constitutes its exercise. Neither is 'contribution' merely an *aspect* of events, given that – in most cases – the specific type of event which 'manifests' a modal property would not happen at all without the ability being exercised. A specific object would not accelerate *at the given rate* it does in the specific conditions if one of the 'contributory forces' were absent; rather, a different sort of event would happen (acceleration at a lower rate and/or in another direction).

Rather than talking about 'contributions' as exercising abilities, we should specify abilities, as I suggested, through their logical connection to counterfactual conditionals. The ascription of abilities entails that a certain type of event would take place if certain, more or less specific, circumstances would occur. In most cases, these circumstances do, of course, include the presence and absence of other objects with some abilities. There is a sense in which these abilities – with their presence or absence – 'contribute' to the manifestation of an event. When their presence is necessary, then the event, *described as a manifestation*, manifests the exercise of these other abilities too. For, as we have seen, the exercise of two (or more) abilities can be manifested by a single qualitative change (or non-change) in certain circumstances.

The coarse-grain description of what happens in a given spatiotemporal region identifies only a qualitative or measurable movement or change. But the proper, fine-grain description – to which science often aspires – identifies two or more events that are going on in the same spatiotemporal region, each being an exercise of different abilities. The resulting state of the sugar's and water's interaction is that *water* is saturated with sugar, *and* at the same time, that *sugar* is dissolved in water. These are two different states of two distinct substances: one is the state of the water, the other is the state of the sugar, and not *vice versa*.

And further, the dissolution of the sugar cube *as* the exercise of the sugar's ability and the dissolving the sugar *as* the exercise of the water's ability cannot be the same events. It is true that when a sugar cube dissolves in water, both the sugar cube's ability and the water's ability are contributing to something. But what they can contribute to is the exercise of each other's ability.

Since this account of events as results of the exercise of abilities implies that two or more events take place at the same spatiotemporal region, one may object that it is an offence against 'economy'. However, philosophical economizing is a delicate and sometimes obscure matter. We should not put much weight on it. It often turns out that we sell cheap and buy very expensive.

Why is it not an offence against economy to say that although only one event takes place in a given spacetime region, it is the result of, or it manifests, different abilities' 'contributions'? We have a rough idea about what events are. We talk about them, we count them, sometimes we observe them. But we rarely talk about, count, not to mention observe, contributions to events. Moreover, although 'contributions to events' are themselves not events, they are assumed to 'manifest' abilities. Whatever is the merit of this idea, it is hard to see how it serves the purpose of 'economy'.

Suppose S throws a punch on W's head, and W faints. It seems that two events have happened because two distinct abilities were exercised: S's ability to punch with a certain force W's head and W's (li)ability to collapse when his head is exposed to a sudden shock. Imagine now, however, that S and W are fighters in a ring. Then that which has happened can be described simply as a knockout. But would our practice to describe a compound of events as a single event cancel out the two component events from the world? This would imply that we can change the ontological structure of the world just by inventing new words: we would have the power to do magic with words.

But this is exactly the case when we describe with one event-word things which happen when two or more objects (or agents), by exercising their respective abilities, interact. Since almost everything that happens in the world involves some such interactions, it is practically entirely justified to classify events that involve frequently reoccurring types of interactions with one word. But classifying events in this way should not mislead us to think that only one thing happens, whereas in fact there are many things going on since many distinct particulars' or substances' abilities are exercised.

It is for this reason that re-describing what happens with reference to such abilities can also explain why a type of event occurred. As we have seen earlier, 'fine-grained' events identified as exercises of different abilities can *explain* why a 'coarse-grained' event (identified as some qualitative change) occurs. Hence, such 'fine-grained' events can hardly be considered as some gratuitous addition to our ontology. The description 'This unsuspended body falls' can be explained as the exercise of the ability of the unsuspended body and as the exercise of the earth's ability to attract another object with a certain mass. That a sugar cube dissolves entirely in my tea can be explained by water at a certain temperature having the ability to dissolve certain sort of substance within a certain time.[26]

Finally, some may object that events cannot be identified as exercises of abilities since the same modal property can, and in fact must, be manifested by several different types of events.[27] Some philosophers even argue that *every* property must have different types of manifestations for this is the 'test of their reality'.[28] Dispositions, as it is often said following Ryle's terminology, can, or must, be 'multitrack'. If abilities too can be associated with many different sorts of manifestations, then events cannot be properly distinguished by the ability which is exercised when they happen.

However, it is hard to make a clear sense of the idea that a modal property is 'multitrack'. To start, we might think that an ability is multitrack if it is manifested by many different types of events. But this is easier to say than to understand. We certainly do not mean by this that some abilities are manifested just by *any* events. Thus, we need to identify something which 'ties' some events together and which explains why any of them counts as the manifestation of a specific modal property. But if there is indeed something which ties them together, then it is *a* more generic ability that is manifested by *a* more generic type of event.[29]

---

[26] Moreover, we can further specify that ability if we add that tea is already acidic water, which can also influence the rate of dissolution.

[27] See, for instance, Molnar (2003), Lowe (2009), McKitrick (2009).

[28] Mellor suggests accepting a principle about the reality of properties, according to which every *real* property must manifest itself in many different ways (Mellor 1974, 175). Interestingly, Ellis and Lierse, who are otherwise sympathetic to Mellor's realist account of dispositions, reject the principle (Ellis and Lierse 1994, 36). However, as I mentioned earlier, they do not think that a kind of process is what it is in virtue of being the manifestation of a modal property. But I cannot quite see how natural kinds of processes can be identified independent of the ability which they manifest. I must add that, in my view, there are not only 'natural' processes and abilities. When one raises a hand in order to signal and when one raises it in order to vote one exercises different abilities since one can have the former without having the latter, and *vice versa*. But these abilities are not 'natural' in the required sense.

[29] Ryle's idea of 'multitrack' dispositions was motivated by his theory of mind, according to which mental states like beliefs and emotions are dispositions, and that there are various different forms of behaviour which can make the same belief, emotion and so on manifest. But, although it is true that

It is interesting to note that Ryle, who introduced the notion of 'multitrack dispositions', says that the concepts which allegedly express such dispositions are *determinables* (Ryle 1949, 44). However, interpreting multitrack dispositions as determinable properties undermines the very idea of 'multitracking'. For if we understand 'multitrack dispositions' as determinable properties, then the (less determinate) *types* of events can always correspond to *one kind* of more specific ability that is exercised.

Since abilities as properties of objects can be more or less generic, the manifestation of a determinable ability can also be a more or less generic type of event. A determinable ability can be exercised in many different determinate ways; just as being red can be manifested by many different specific shades. But an event that manifests the more determinate ability is also a manifestation of the ability which is its determinable. Thus, the one-to-one correspondence between kind of abilities and types of events which is their exercise does not break down.

However, I must note that even if this parallel with colour-properties helps to illustrate the connection between determinate properties and their determinable, it might be misleading in another respect. The case of colour suggests that knowing the determinate-determinable connection is a purely *a priori* matter. But there is no reason to believe that it always is. Oftentimes, it is a matter of theoretical discovery which determinate falls under some determinable.[30]

That the free fall of a body and the orbit of the Earth both fall under the determinable 'manifestation of the ability of objects with a certain mass to attract others', or that different forms of symptoms are manifestations of the same kind of bacterial infection, is certainly not just an *a priori* matter. They can be the subject of important scientific discovery. But the possibility of such discoveries does not undermine the philosophical point that events can be classified as exercises of abilities. Some abilities are determinates of some determinable, as are the corresponding events which are their manifestations.

---

one's belief that *p* can explain rather different sorts of behaviour, what this shows is that we might not want to interpret beliefs as dispositions after all. Acquiring certain beliefs can, indeed, explain different behavioural abilities and dispositions, but beliefs *themselves* might not be dispositions at all.

[30] As Stephen Yablo notes, just as identity, the relation between determinates and their determinable can be empirically discovered and yet metaphysically necessary. See Yablo (1992). For further important discussions about the determinable-determinate relation involving modal properties see Wilson (2021).

7

Possibility, actuality and worlds

For to substitute the logical possibility of the concept *(namely that the concept does not contradict itself) for the transcendental possibility of* things *(namely, that an object corresponds to the concept) can deceive and leave satisfied only the simple-minded.*
(Kant *The Critique of Pure Reason*, B 302)

Essential to science is not so much that a pure immediacy should be the beginning, but that the whole of science is in itself a circle in which the first becomes also the last, and the last also the first.
(Hegel *The Science of Logic*, 21.57)

My main concern in this book is to understand the nature of contingency in our world. I have argued that we cannot understand contingency's nature without a proper account of modal properties. In fact, a theory of contingency *is* essentially an account of modal properties. Hence, I cannot complete my project without addressing a challenge raised against the very idea of there being first-order modal properties in the world.

As I noted in the introduction, the ability-based account of contingency is meant to provide an alternative to the standard possible world accounts. Possible-world accounts of contingency are not logically incompatible with the ascription of first-order modal properties. Nonetheless, these accounts of contingency typically deny that there are first-order modal properties. In fact, even a stronger claim seems to be true: the world-account of contingency is at least partly motivated by the belief that modal properties cannot be fundamental.

Those who believe that the contingency of events in our world is grounded in first-order modal properties need not deny that there are possible worlds. True, many philosophers aim to understand contingency with reference to modal properties precisely because they are sceptical about the very idea of possible worlds as truth-makers of modal truth. But there is no inconsistency in holding the view that, while contingent possibilities *in* this world are grounded in the distribution of modal properties in it, there might be other possible worlds with modal properties that are different from the ones instantiated in our world.

In fact, there is no strong reason to reject the idea of 'other' (that is to say, nonactual) possible worlds.[1] Denying that there are such worlds implies that our *whole* world, with all of its modal properties (capacities, physical properties, dispositions and abilities) in it, is not contingent. And, as far I can tell, we do not have sufficient reason to deny that there might be other worlds with different modal properties in them. For it seems, at least to me, that it is an entirely contingent matter that this world, that is to say, *this* totality of particulars with *these* modal properties in *this* spatiotemporal manifold, exists.[2]

If the existence of objects and the occurrence of events depend on the distribution of modal properties in a world, then the fundamental modal structure of a world must be characterized with the help of the distribution of modal properties. This means that if the distribution of modal properties were different than it actually is, then different contingencies would exist in the world: different kinds of events could and would happen, and different sorts of objects would exist.

Thus, nothing for which I have been arguing here is incompatible with the view that there are other possible worlds. Neither do I want to commit myself to their existence or about any specific account of their nature. All I want to say is that I am not aware of any reason that would be sufficient to reject the idea of possible worlds. What we do have sufficient reason to reject, however, is the view that the contingency of things and events *in* our world can be understood with reference to other, from our perspective nonactual, *worlds*. Hence, as far as the property-based account of contingency goes, the important question is not why we should reject possible worlds. What we need to understand is why so many philosophers reject the fundamental role of first-order modal properties in this world.

The main reason seems to be some considerations about the notion of actuality. In *Metaphysics* ix Aristotle claims that *energeia* or *entelecheia* is prior 'in

---

[1] There is, of course, a further question about how to understand the nature of the nonactual worlds, that is, whether we should consider them as entities of the same kind as our own world is. I do not have much to add to the discussion of this issue, but luckily, it is not relevant to my present topic.
[2] Surely, this is a controversial claim. Some important philosophers, Hegel included, thought that it is not possible that any other world than the actual exist. Their arguments for the impossibility of other worlds are based on the principle that whatever exists, including the whole world, must have a sufficient reason for its existence. I cannot discuss the plausibility of this this principle within the confines of the present enquiry.

time', 'in account' and 'in substance' to *dunamis*.³ If *energeia* and *entelecheia* are interpreted, roughly, as the actual exercise and the actual (completed) products of the exercise of modal properties (*dunamis*), then Aristotle's claim should be understood as the thesis that the actual is, in three distinct ways, prior to the possible. Must then those who are committed to the fundamentality of modal properties reject the priority of actual? And if so, does it follow that everything there is, is 'mere potentiality'?

Not at all. In fact, as I shall argue in this chapter, the priority of the actual is almost a direct consequence of the modal property approach to contingent possibilities. If possibilities are grounded in the modal properties of objects and substances, then those properties and their bearers must actually exist in order for anything to exist as a possibility. Contingency *in* a world presupposes that there is a world in which the modal properties are instantiated, and which is prior 'in time', 'in account' and 'in substance' to any possibility in it.

Crucially, however, in contemporary debates, 'the priority of the actual' is not interpreted as the ontological priority of matter, the individual objects which are constituted by it, and the existence of modal properties instantiated by the latter, to contingent possibilities in the world, but rather as a thesis about the fundamentality of the 'categorical' or 'qualitative'. If the priority of actual meant that fundamental properties are 'categorical' or 'qualitative', then *first-order* modal properties could not exist. The distribution of modal properties should somehow be derivative of the distribution of assumedly nonmodal, 'categorical' or 'qualities', properties.

In this final chapter, I wish to round off my discussion about the role of modal properties in ontology by explaining why I find (a) the idea of 'fundamental nonmodal properties' entirely empty and, correspondingly, (b) the rejection of the existence of first-order modal properties unjustified. And since the thesis about the priority of the nonmodal is logically connected to the world-account of contingency, if the latter needs to be rejected, then we have a further reason to believe in first-order modal properties.

In what follows, first, I shall offer a hypothesis about the genealogy of the world-based account of contingency which explains how naturally it leads to the assumption that fundamental properties must be 'categorical' or 'qualitative'. Correspondingly, I shall explain why the idea of a fundamental 'categorical' structure of reality is empty. All this does not mean that there are not any 'categorical', that is nonmodal, properties in the world. What it means is only

---

³ See Aristotle *Metaphysics*, ix, 8, 1049b–50b.

that those properties cannot be fundamental. I shall conclude the chapter with some considerations about how the empiricist view which interprets sensible qualities as nonmodal and fundamental presupposes a form of idealism.

From events to worlds

Many sorts of things can be contingent: the occurrence of events, the existence of persisting particulars, the distribution of properties and even the whole world. There is sense, however, in which the modal property-based account of contingency assigns a special role to events and processes in the understanding of contingency in a world. Contingency is understood here as a *dynamical* phenomenon. Persisting substances are contingent because they can come about and pass away and hence their existence depends on their 'birth' and 'death', which are special sorts of events. Hence, the existence of contingent objects and persons depends on the exercise of modal properties which makes their coming to existence possible.

In another sense, however, the modal property-based account does not assign ontological primacy to events *in general*. For it explains what happens or exists contingently by ascribing modal properties either to some persisting particular, physical objects or persons; or to some samples of chemical substances; or to some ensemble of objects or persons. What can and cannot *happen* depends on what objects, persons or substances can or cannot do: some cups can break; therefore, there must be a sense in which it is possible that they break even if, actually, they never do. Some people can win an election; therefore, there must be a sense in which it is possible for them to win, even if, actually, they lose.

As mentioned, most contemporary accounts of contingency understand what objects or persons can do, even if they do not, in terms of the contingency of our world. But why should the contingency of certain events *in* our world be understood with reference to a whole world? Here is a simple, and at least *prima facie* appealing, explanation given by David Lewis.

> I believe that there are possible worlds other than the one we happen to inhabit. If an argument is wanted, it is this. It is uncontroversially true, that things might be otherwise than they are. I believe, and so do you, that things could have been different in countless ways. But what does this mean? Ordinary language permits the paraphrase: there are many ways things could have been besides the way they actually are. On the face of it, this sentence is an existential quantification. It says that there exist many entities, of a certain description, to wit 'ways things

could have been'. I believe that things could have been different in countless ways; I believe permissible paraphrases of what I believe; taking the paraphrase at its face value, I therefore believe in the existence of entities that might be called 'ways things could have been'. I prefer to call them possible worlds. (Lewis 1973, 84)

Lewis's aim in this passage is to defend a specific vision of the ontological status of nonactual worlds, according to which they are concrete, but causally and spatiotemporally separated, universes. This vision, as Lewis himself says, 'met with many incredulous stares'.[4] But how we interpret the notion of possible worlds is one question; it is another whether claims about possibilities – the 'ways things could have been' – must be understood as a claim about *worlds*. Many philosophers, who disagree with Lewis about 'the ontological status of worlds', still agree with him that the 'ways *things* could be' must be understood as *worlds*.

We might call Lewis's argument for this view the *natural paraphrase* argument. Its first step is the idea that when we talk about contingencies, we *quantify over* certain entities. When we say that some particular thing *can or could* be otherwise,[5] that is, when we express possibilities by using some modal auxiliaries, what we really mean is that '*there are* ways such that . . .'. When we think about possibilities, we in fact quantify over something.

The second step of the argument is the claim that *what* we quantify over ('the ways things might be') must be understood as possible *worlds*. When I say that Jack could be grown taller than he actually is, I do not just mean that Jack has possibly grown taller or that it is possible *for him* to have been grown taller. Rather, what I mean is that *there is another world* at or in which he is taller. Debates in modal metaphysics are replete with discussions about whether Jack is the same person in the actual and the possible cases; that is, how to make sense of 'his counterfactual identity'. The question why we need to refer to whole worlds in order to understand this possibility at all has received less attention.

According to the world-approach, when we say that '*p* can or could be the case'; or '*p* is possibly the case'; or 'possibly, it is the case that *p*'; that is, when we modally modify a statement, then, implicitly, we use a quantificational language. For this reason, when we say that something *possibly* happens or exists, then we

---

[4] Lewis (1973, 86; 1986, 133).
[5] Lewis uses 'might', but I try to avoid this, because, as I argued in Chapter 5, 'might' expresses epistemic possibilities, while my concern here, as Lewis's, is metaphysical possibility.

must be committed to the view that *there is a possibility* that that thing happens or exists.

It is important to note that this is not at all a self-evident or trivial claim. Suppose that something *has to* happen or exist in the sense that it *necessarily* happens or exist. 'Necessarily' modally modify the simpler proposition that something happens or exists. But we would not naturally paraphrase the modal sentence like 'there is a necessity' that this thing exists or happens. This would be a rather bizarre way of speaking, if not straightforwardly incomprehensible. And further, we would not introduce into our ontology 'necessary worlds' in order to explain what 'J necessarily φs' means.

We might of course try to understand 'J necessarily φs' as 'J φs in every possible world'; or as 'In every possible world, it is true that "J φs"'. But this is a *theory* of necessity, not a natural paraphrase. People who understand 'J necessarily φs' need not implicitly understand it as J's φ-ing in every possible world. In fact, most people who understand the first sentence do not at all understand the second until they have learned some modal logic.

In contrast, it is often taken to be self-evident that when we modally modify a sentence as 'J can φ' or 'J possibly φs' we quantify over possible events or objects, and that this quantification involves possible worlds. What I try to stress is that this 'paraphrase' is a heavy-loaded philosophical theory, which is not simply entailed by our ordinary capacity to understand statements about possibilities. Just as understanding necessity as truth in or at all possible worlds is a *theory* about necessity and not a simple paraphrase, understanding possibilities as truth in or at a world is a theory about possibility which is not a mere consequence of the use of those statements that express claims about contingencies. The reference to possible worlds involves a theoretical commitment driven by a metaphysical theory of the ontological ground of contingency.

Hence, the paraphrase which Lewis introduces does not seem to be so natural after all. It is natural only from the perspective of a specific theory. That theory assumes that only what happens in other worlds can *represent* adequately contingencies (and also necessities) *in* our world. This may or may not be right, but we still do not know what *the ontological ground* of contingency, which those representations allegedly represent, is. Next, I shall address briefly Lewis's response to that crucial question, since this helps explain the connection between our main concerns in this chapter: the world-based account of possibility and some philosophers' scepticism about first-order modal properties.

Contingency and contradiction

The notion of possible world was introduced in the seventeenth century by Leibniz. He introduced the concept in order to solve a specific metaphysical and theological problem: the problem of theodicy. Leibniz suggested that 'the problem of evil' can be answered if we concede that God must have created the world which actualizes the best possible balance between things that are good and things that are bad (or evil). There is no possible world with *more* goods and *less* 'evils' in it than the actual world. In this sense, our world is the best of all possible worlds. Hence God's infinite power and benevolence are compatible with the actual existence of evil.

This response to the problem of evil requires a comparison of whole worlds. From the perspective of singular events, we cannot understand the *reason why* God allows that bad or evil things happen. From the global perspective, which we humans with our limited epistemic capacities could not comprehend, everything happens in order to make room, as it were, for the most possible goods.

But, importantly, Leibniz did not use the notion of possible worlds to explain contingencies *in* the actual world.[6] In fact, it is contentious whether there could be contingencies in Leibniz's philosophy at all. Perhaps, Leibniz was a 'determinist' in the sense of being a necessitarian.[7] Nonetheless, there is one aspect of Leibniz's account of contingency which did exercise a significant influence on the later interpretations of contingency in terms of possible worlds. This is Leibniz's claim that something is impossible if it involves a contradiction. Otherwise, it is a possibility. That is to say, something is possible only if it does not involve a contradiction.[8]

---

[6] For an excellent discussion about the relation between Leibniz's view of necessity and that of modern modal logic as truth in all possible worlds, see Adams (1994, 46–50). For a conflicting opinion, see Mates (1986, 107). Mates claims that there is always 'the presumption . . . that a necessary truth is a proposition true of all possible worlds', but he admits that it is 'almost never stated explicitly'.

[7] Leibniz himself claims on several occasions that 'determinism does not entail necessity'. For a defence of Leibniz's distinction in modern context see Huoranszki (2011), Chapter 2. For interpretations of Leibniz as a 'determinist' in the sense of being a necessitarian see Adams (1994), Part 1 and Griffin (2013).

[8] Leibniz was committed to this doctrine throughout his carrier, from at least the 'Discourse on Metaphysics' (1686) XIII to his 'Monadology' (1714) 33. About the context and the history of the noncontradiction-interpretation of possibility see Wilson (1990), Chapter 1. It is interesting to note that later Kant became critical to this account of possibility because, as he says, 'It is, indeed, a necessary logical condition that a concept of the possible must not contain any contradiction; but this is not by any means sufficient to determine the objective reality of the concept, that is, the possibility of such an object as is thought through the concept.' Kant (1929/1781, 240) (A 220, B 268)

However, we cannot make much sense of the idea that the occurrence of *an event* or the existence of *a substance* (understood as a persisting particular) 'involves a contradiction'. What could it mean that an event or a substance is 'inconsistent'? In order to apply the criterion of noncontradiction, we need something that can be true or false: assertive sentences, statements; or propositions which such sentences or statements express.

Further, if contingency is understood as noncontradiction, it must be understood with reference to some *sets* of statements or propositions. One may object that particular statements can involve a contradiction if they state a 'conceptual impossibility'. It is a 'conceptual impossibility' that Jack is a married bachelor. (Whether or not such sentences can express a proposition is a further question, which depends on one's specific account of propositions. I shall not pursue this issue further.)

But it is important to note that even 'conceptual impossibility' must be a *consequence* of the inconsistency of some distinct propositions. 'John (or someone) is a married bachelor' is a 'conceptual impossibility' because the two propositions 'John (or someone) is a bachelor' and 'John (or the same person) is married' cannot be both true at the same time. 'Conceptual impossibility' is always a consequence of the contradiction between two or more propositions.

The concepts 'round' and 'square' themselves do not entail any 'conceptual impossibility'. For it is perfectly possible that, in the very same world, figure A is round and figure B is square. It makes no sense to claim that these concepts themselves are contradictory. What is indeed impossible is that figure C be round, *and* C be square, in the same world at the same time. That's what it means to say that 'round square' is a conceptual impossibility. But the impossibility of round square does not follow from the concepts themselves, but from the *propositions* which ascribe contrary properties to the *same* object at the same time.

This has an important consequence as to how to interpret the notion of contingency. Suppose that a certain event *e* is not part of the actual history of this world. If this is only a contingent fact, then event *e* could have happened. According to the noncontradiction account, *e* is possible if the proposition '*e* occurs at t' is not contradictory. The proposition that '*e* occurs at t' should be consistent – but consistent with what? The natural response seems to be that it should not contradict *anything*. And this means that what we need to consider when our aim is to understand whether or not *e*'s non-occurrence was contingent is whether or not the truth of '*e* occurs at t' is consistent with every other truth in or about a world.

It is the consequence of the 'noncontradiction-interpretation' of possibility that, more generally, contingency is often understood in terms of maximal consistency. An event could occur, or a persistent particular could exist, if its occurrence or existence is compatible with everything else. More precisely, an event could occur, or a particular could exist, if *there exists* some consistent and *maximal set* of propositions, which includes the proposition that the event occurs or that the particular exists. This means that possibility is, by its nature, a holistic concept. The possibility of a specific event or a particular object presupposes a consistent set of propositions of which the given proposition is an element.

This holistic interpretation of possibility has turned out to be very useful in modal logic, that is, in the study of correct modal inferences. The assumption that what is possibly true is true at least in one possible world and what is necessarily true is true in all possible worlds, together with various assumptions about the 'accessibility' of these worlds, has turned out to be an excellent device in model theoretic semantics. It also helped to develop various alternative systems of modal logic and uncover the logical relations among them.

Noncontradiction is arguably a good criterion of *logical* possibility. According to the criterion of noncontradiction, proposition p is *logically* possible, if p is a member of a maximal set of consistent propositions. However, I shall argue that noncontradiction is not helpful when we want to understand *contingent* possibilities. By 'contingent possibilities' I mean possibilities that are not *merely* mathematically or logically possible.

The square root of 4 *can* be –2 or +2. When 'If p then q' is true and p is false, then q *can* be true or *can* be false. These are possibilities, but not contingent possibilities. Certainly, if the proposition 'e can happen' or 's can exist' is true, then e's happening or s's existence cannot be logically impossible. But this does not explain *why* the occurrence of e or the existence of s is possible. Contingency, as it is often said, is 'restricted possibility'; and the restrictions we have in mind are not only restrictions of the mathematical or logical kind.

In fact, many contemporary accounts of contingency would agree with this. But they would still raise the question of possibilities as a question about possible worlds. There are some interesting reasons for this, which go beyond the idea that noncontradiction or consistency is the only constraint on possibility. In the following section I explain what I think those deeper reasons are and why I do not think that they justify the use of worlds in an adequate account of contingencies in a world. We shall also see how the confusing idea that the 'actual' must be 'categorical' or 'qualitative' arises from this account of possibility.

Recombination and independence

In many contemporary accounts of possibility, the ultimate ground of contingency is not noncontradiction *simpliciter*, but noncontradiction together with the 'principle of recombination'. Outside the scope of logical possibility, it is this latter principle that is supposed to explain why certain propositions are or are not consistent.

The principle of recombination is taken to be a consequence of 'Hume's dictum', according to which there are no necessary connections between distinct existences. Unfortunately, in this form, 'Hume's dictum' is not very helpful. Everyone should agree that there are not necessary connections between distinct existences, if what it *means* that one thing is metaphysically distinct from another is that there is no necessary connection between them. We still need to understand what makes things distinct.[9]

In my view, the so-called Humean theories of contingency have no plausible answer to that question. Merely on the basis of 'Hume's dictum' we cannot *argue* that some things are possible whereas others are not; and we cannot decide which propositions can be 'recombined' with each other without involving a contradiction. According to Humeans, possibilities as possible worlds should be understood in terms of different combinations of *independent* nonmodal properties. But independence *presupposes*, rather than explains, contingency.

To see this, compare the issue about properties' independence *of each other* to the issue about particulars' or individuals' independence of their properties and/or of other individuals. Some essentialists believe that 'essence precedes existence' meaning, roughly, that the existence of an individual is not independent of which properties it instantiates. More specifically, an individual can exist only as an instance of some (essential) kind.[10] Other essentialists believe that the identity of an object or an organism is not independent of some other individuals: what or who an individual is depends on their causal or genetic ancestors. Similarly, properties are not independent of each other if the instantiation of one of them (by some object(s) at a spatiotemporal location) is (metaphysically) determined by the instantiation of another.

In contrast, properties P and Q are *in*dependent if the instantiation of property P by a particular does not logically and/or metaphysically entail the instantiation of property Q either by the same particular at the same time; or by the same

---

[9] For more on this issue, from a different perspective, see Wilson (2015).
[10] This is not the only interpretation of this principle since, on some views, the essence of a particular is not a property.

particular at another time; or by any other particular at any other time. Thus, being red and being coloured are not independent, since if something is red then it is necessarily coloured; but being red and being round are independent. Similarly, 'being taller than' and 'being shorter than' are not metaphysically independent properties, since if Jack is taller that Jill than it follows that Jill is shorter than Jack. And if Jack is older than Jill, then Jill was born later than Jack. But if Jack was born south of London, it does not follow that Jill was born north of London (or, in fact, at any specific place).

Thus, the applicability of the Humean principle of recombination presupposes that we can distinguish those properties the instantiations of which are essentially or logically connected from those which are not. Contingent possibilities are explained then with reference to worlds in which properties that are *not* necessarily or logically connected are distributed otherwise than they actually are.

But such 'Humean' theories must face a fundamental difficulty. For if independence is assumed to be the ground of free recombination, then independence *must be* a modal notion. After all, two properties are independent in the sense of not being essentially or logically connected only if they are freely recombinable. But the free recombin*ability* of properties, as the very word suggests, is a (higher-order) modal property.

But Humean accounts of modality aim to understand contingency *without* reference to modal properties. They use worlds rather than properties for explaining contingency because they understand possibilities as combinations of logically and metaphysically independent properties, and the recombinability of all relevant properties are best represented by worlds. However, whether or not two properties are freely recombinable must somehow follow from the nature of those properties. Thus, it seems difficult to make sense of the idea of independence without already assuming that the relevant properties *can* be recombined.

Hence, the ultimate difference between Humean theories of contingency and their alternatives is their rival views of the nature of properties. For the Humeans, the ground of contingency is that the instantiation of certain properties does not constrain the instantiation of other properties. As we have seen, this *is* a modal characterization of properties. But Humean theories seem to assume that the modal nature of properties can be revealed only by their interconnections. The mere instantiation of a single modal property by some object(s) at one spatiotemporal region cannot explain contingency.

I do not pretend that I have a knockdown argument against the Humean idea that the ultimate ground of contingency is the recombination of nonessentially or nonlogically connected properties. To use Hugh Mellor's memorable metaphor, philosophy is often like boxing. We often cannot settle issues by a knockout, so we need to settle them on points.[11] The point I wish to score here is based on the observation that, *if* Humean accounts are motivated – as they often seem to be – by the attempt to give a reductive analysis of modality in terms of nonmodal facts and properties, *then* they seem to be doomed. Since independence is obviously a modal constraint, Humean recombination cannot provide a reductive analysis of modality. So at least one reason the reject first-order modal properties is unjustified.

However, in my view, the problems with this approach go deeper than failing to satisfy some standards of reduction. In fact, as I have argued in Chapter 5, the idea of a *non*reductive analysis of modal properties makes good sense. But the fundamental difficulty with a Humean analysis of contingency is not that it fails to provide a reductive analysis. The fundamental difficulty with it is that it does not seem to provide any.

Consider the following claims. Donkeys as we know them do not talk. Is this a contingent fact? Could there be talking donkeys? Electrons as we know them are negatively charged particles. But is this a contingent truth? Could some electrons be positively charged? If a particle is negatively charged, it will be disposed to repel other negatively charged particles. Is this a contingent fact? Can negatively charged particles be disposed to attract other negatively charged particles?

Some may think that all these are genuine possibilities. Others, like me, may disagree. But whatever one's stance on this issue is, the principle of recombination is not going to provide any guidance about how to resolve these questions.

Consider again the example of talking donkeys. Those who think that donkeys can*not* talk will argue that being a donkey and being a talking animal are *not* freely recombinable properties. Certainly, one can conceive or even imagine that an animal, which *looks* like a donkey, talks. But this hardly proves that there could be talking donkeys. Perhaps, all this shows is that there could be talking tonkeys: animals that are similar in many respects to donkeys *and* talk.

To see why this proposal makes good sense, consider the following question: can some donkeys reproduce by laying eggs? This possibility does not seem to be more distant than the possibility of talking donkeys. Someone who wants to deny that this is a genuine possibility could argue that, indeed, we might conceive or imagine some animal which looks in many respects like donkeys do, but it is *obvious* that if they reproduce by laying eggs, then they are *not* donkeys.

---

[11] See Mellor (2004, 309).

For donkeys are, essentially and hence necessarily, mammals and mammals are, essentially and hence necessarily, viviparous. Being a donkey and being oviparous are not independent properties because they cannot be co-instantiated by the same individual.

The Humean might respond that this dispute is merely verbal. For being a donkey and being a talking animal are not 'fundamental' properties. On that view, such properties supervene on the distribution of more fundamental ones. Whether or not there could be talking donkeys depends on how those more fundamental properties can be distributed. If they can be distributed in a way that gives rise to talking donkeys or donkeys which lay eggs, then it is only a contingent fact that donkeys in our world do not talk.

Someone who rejects such possibilities can always insist that if 'being a donkey' is a genuine property, then that property is *not* recombinable with being a talking animal or being oviparous; and this will not be altered by the fact that there might be some *other* properties (like being a tonkey) which are. Yet, the Humean can insist that if we do not want to *call* the donkey-looking talking animal a donkey, that's fine. But it is a contingent fact that we do not meet such animals in our world. In another world, in which fundamental properties are arranged otherwise, *tonkeys* can exist, and this is all that we need in order to understand contingency.

But invoking the idea of 'fundamental properties' does not seem to resolve the question whether donkeys (or tonkeys) can talk or reproduce by laying eggs. It just pushes the problem one level down, as it were. For we still need to show how the fundamental properties can be recombined so that this will give rise to talking donkeys (or tonkeys, never mind the difference). It is, after all, the independence of these assumedly fundamental properties which ground the independence of all other properties. Thus, we should at least clearly understand what the independence of those properties means. The Humean, unfortunately, has no answer to this problem.

Donkeys, like every other macroscopic entity, are composed of microphysical particles. Among those particles some are electrons. Suppose that electrons are fundamental particles. (It is, of course, logically and metaphysically possible that electrons themselves are constituted by some even more fundamental entities. But since everything I shall say applies to those entities as well, this does not matter.) Such particles are identified by some of their assumedly fundamental properties: their charge and mass, for instance. The idea is then that these properties are independent and can be freely recombined in the sense discussed earlier.

However, the question of contingency reappears. Are such 'fundamental' properties indeed freely recombinable? Normally, we identify such properties with reference to how 'fundamental particles' instantiating them move and interact. For instance, a particle having a certain electric charge is disposed to attract particles with the opposite charge and is disposed to repel particles with the same charge. But this seems to imply that the instantiation of a fundamental property like electric charge does have metaphysical implications about what other properties will be instantiated at other spatiotemporal regions.

Similarly, the way scientists normally identify such properties seems to imply that some fundamental properties cannot be co-instantiated at the same spatiotemporal region. Perhaps the instantiation of a determinate mass at a certain spatiotemporal region does not entail which charge is instantiated there. But negative and positive charge cannot be instantiated at the same spatiotemporal region, even if being negatively charged and being positively charged are distinct fundamental properties. Some fundamental properties do not seem to be 'freely recombinable'.

This is a much-discussed problem in philosophy of science, which convinced many philosophers that fundamental physical properties are modal properties after all.[12] And if the distribution of fundamental properties is modally constrained, then the distribution of the 'less fundamental' resultant properties must also be so constrained. Thus, it might not be a contingent fact after all that there are not talking donkeys; or that there are not tonkeys.

Humeans can, of course, insist that it is only a contingent fact that positive and negative charge cannot be instantiated at the same spatiotemporal region. They might complain that the arguments against such possibilities confuse the way in which we identify fundamental properties with what those properties really are. It is true that we identify such properties with reference to how particles instantiating them would typically behave in certain circumstances. Thus, we identify them by their role in explaining certain interactions. But this does not mean that such roles tell us what those properties 'intrinsically' are.

Put otherwise, Humeans do not deny that there are *second-order* modal properties which are *not* independent of each other or of what happens in the world. Their claim is that such properties do not explain what can or cannot happen. If someone is a murderer, then it is necessary that he killed someone else because being a murderer just means that the murderer intentionally killed someone. But this does not mean that it is not a contingent fact that Jack the

---

[12] For further exploration of this problem and its consequence see Huoranszki (2019).

Ripper, who actually killed many persons, is a murderer. In another world, he might have only done *postmortem* autopsy.

Analogously, in *our* world certain fundamental properties, given the patterns of their instantiations, give rise to certain distribution of second order 'role' properties. Given this distribution, donkeys are viviparous; hence, they can reproduce only in a certain way and they do not (actually) lay eggs. But given that fundamental properties are freely recombinable, the patterns of their instantiation in other worlds can give rise to other roles, and hence other resulting nonfundamental properties. There might be then donkeys (or tonkeys) which do lay eggs.

Now, the postulation of this type of first-order nonmodal properties seems to me entirely ad hoc. But, as far as our understanding of contingency is concerned, the problem goes deeper. For these properties are postulated, at least partly, in order to ground contingency at the 'higher level'. Yet, it is hard to see how they can fulfil that role.

On the one hand, the content of the relevant properties can never be known to us. We can identify only the second-order 'role properties', but we are bound to remain – as Lewis says perhaps echoing Locke's famous phrase about 'real essences' – 'incurable ignorant' about their nature.[13] But, on the other hand, possibilities in our world are supposed to be grounded in the independence of such fundamental properties and hence their free recombinability. Thus, although we have no clue regarding what the properties that explain contingency are, we introduce an entirely a *priori* modal condition that they need to satisfy, namely that they are (must be?) freely recombinable.

Moreover, we do not have the slightest idea about how the independence of the postulated properties ground other claims about possibilities which we do seem to understand perfectly well; like the claim that there could (or could not) be talking donkeys. Since being a donkey is not a fundamental property, it must be 'a higher order role property'; in other words, something is a donkey if it does what donkeys typically do. Donkeys typically heehaw; they do not typically talk. But then, it seems that no donkey can talk. We have seen earlier, however, that tonkeys might talk, for tonkeys are animals similar in many respects to donkeys, except that they sometimes talk. But can tonkeys exist?

They can if the fundamental properties can be recombined in a way which results in the existence of some tonkeys. But we have no clue whatsoever about whether they can indeed be so recombined. Even if we take conceivability to be

---

[13] See Lewis (2009).

our guide of possibility, it can certainly not be our guide about the compossibility of such properties the content of which we are bound to be entirely ignorant. And we do not have any other guide either. Thus, ultimately, the ground of metaphysical contingency is the distribution of such properties which are postulated, at least partly, to explain this contingency, but the nature of which is bound to remain beyond our comprehension.

I am not arguing that the Humean theory of contingency is thereby conclusively rejected. The real problem seems to me that we have no idea about how it could be rejected. According to such theories, contingency is grounded in the principle of recombination, but we cannot understand what it means that the properties which assumedly ground possibilities are 'freely recombinable'; all we can know is that, by hypothesis, they must satisfy the requirement of independence. And the only reason to postulate such recombinable properties is that they can explain contingency, which is the very issue. This seems to me a sufficient reason to endorse an alternative account of contingencies in our world.[14]

Essentialism: A world without contingency

Standard possible world analyses of contingency presuppose the existence of 'fundamental natural qualities', the recombinations of which are claimed to constitute distinct worlds. I argued that it is impossible to identify, and hence comprehend, such qualities. This complaint is not entirely new.[15] If the instantiation of a property at a moment has no logical or metaphysical implications about what can happen or exist in the world then or at other times and hence what other properties can be instantiated, then it seems to be impossible to identify that property.

I certainly agree with this complaint to the extent that the recombination account of possibility fails to explain how we can identify the properties that we are supposed to recombine. However, my account of the modal nature of properties differs fundamentally from most other philosophers' view who also reject the Lewis-type account of properties. According to the account I endorse, modal properties entail what *can* happen in the world. According to the most

---

[14] For further problems about the Humean account of contingency especially in the context of physical laws, see, again, Huoranszki (2019).
[15] Following Shoemaker (1980), though Shoemaker's arguments are very different from mines and are unrelated to the issue of possible worlds. For further arguments see also Bird (2007), Chapter 4.

popular, alternative response, the instantiation of properties entails what *has to*, by metaphysical necessity, happen in the world. We might call this latter view – following two of his most important representatives – either 'scientific essentialism' or 'dispositional essentialism'.[16] Whenever I talk about 'essentialism' in what follows, I have this version of essentialism in mind rather than traditional 'metaphysical' essentialism.

While traditional essentialism was primarily concerned with the nature of individual substances and only derivatively with essences of natural kinds which may or may not have been understood as some sort of properties, scientific or dispositional essentialism is a general outlook about the nature of properties themselves. Which class of properties are, according to the essentialist view, modal in this sense is a contentious issue. Some hold that only the 'fundamental physical properties' are modal in this sense. Others argue for this type of essentialism mainly by trying to understand the nature of sensible properties of macroscopic objects.[17]

Although, as it should be obvious by now, I agree with the idea that (most, even if not all) properties' instantiation must have implications as to what happens in the world, I doubt that essentialism provides an adequate account of properties or of their modality. First, some versions of essentialism are ultimately a reinterpretation of Lewis's view about 'roles' and their 'fulfillers'. But while we can clearly make sense of Lewis's distinction between functionally identified *roles* in contrast to the *properties* which are supposed to 'realize' them even if we find the nature of the latter inscrutable, it is unclear to me how we could make clear sense of the relation between *properties* and the powers which *they* are supposed to possess. Second, and more importantly, the essentialist account holds that properties *necessitate* events in the world; and I doubt that there is any route from such necessities to the explanation of contingency in a world.

As to the first problem, it seems that dispositional essentialist accounts of properties try to *reinterpret*, rather than reject, Lewis's distinction between 'roles' and 'role-fulfillers'. As we have seen, the modally involved properties have, for Lewis, only a derivative function in ontology: their function is only to help *us* to identify *a role* for some properties, which is then contingently 'fulfilled' by the

---

[16] After Ellis (1999) and Bird (2007).
[17] As to the first type of essentialism, see Bird (2007) and especially Bird (2016), where Bird argues that many nonfundamental properties are not modal. See also Swoyer (1982) and Bigelow, Ellis and Lierse (1992). For the second, see Shoemaker op. cit., in which he uses 'being knife shaped' as his major example. Ellis holds that the applicability of essentialist thinking depends on the type of science we are interested in. For instance, physical and chemical kinds are apt for an essentialist analysis; biological species are not. See Ellis (1999, 169–71).

ontologically fundamental 'qualities'. The essentialists object that properties – or at least those properties with which they are especially concerned with – cannot fulfil their ontological role unless they have *powers* essentially.[18]

However, these essentialists are still concerned with how to identify distinct properties across different possible worlds. Unlike Lewis they hold that we can identify them only if we associate properties with such causal powers which entail what other properties occur in the world when they are instantiated. For instance, the property 'negative charge e^-' is individuated by its power to attract things with some opposite charge, while it also has the power to repel things with the same charge. A property has many such 'individuating' powers, which we may call the property's 'causal profile'.[19] But what then are these causal powers?

One possible answer is that they are some sort of second-order properties of properties. However – apart from the technical problem that this answer leads to an inadmissible regress similar to the one that I have discussed in Chapter 1 – it just seems wrong to ascribe powers (or potencies) to properties. Ordinarily, we ascribe powers to persons and objects. We 'empower' or 'disempower' them, not their properties.[20] Redness, which is a first-order property of some objects, has the property of being a colour as opposed to being a voice. Courage, which is a first-order property of some persons, has the second-order property of being a virtue as opposed to being a vice. But what does it mean that the power to attract objects with positive charge is a property of negative charge? We attribute this power to an electron, if anything, and not to its being negatively charged.

This implausible account of powers seems to me a consequence of the way dispositional essentialists approach to the problem of modal properties. They approach the question about the *identity* of modal properties as if it were analogous to the problem of *individuation* of particulars. The assumption here is that the Lewis-type and Armstrong-type 'qualities' must have 'quiddities', which are supposed to fulfil a role analogous to 'haecceities'.[21] But I find this analogy rather misleading. Haecceities are *properties* that individuate *particulars*. They are *non-qualitative* individuating properties as, for instance, the property of 'being identical with Hume' is. 'Quiddities', if they mean anything, would be such (higher-order) properties the instantiation of which would make a property the

---

[18] The terminology, as usual, is not unified. Bird talks about 'potencies' rather than causal powers. I do not think, however, that anything in substance depends on the choice of words here.
[19] I borrow the terms from Hawthorne (2001).
[20] This problem was noted by Menzies (2009).
[21] See again Hawthorne (2001) and Bird (2007), Chapter 4.

quality that it is. But they cannot 'individuate' any property, for properties are not individuals.

Dispositional essentialists, of course, reject 'quiddities'. Yet, some of them seem to grant the assumption that we must find a criterion which helps us to identify properties across possible worlds as we need to find a criterion with the help of which we identify individuals (or their counterparts) in this context. But there is no such criterion, for if there are distinct possible worlds, then the properties in those worlds must be different from the properties in the actual world. This seems to be the consequence of any theory according to which properties are *modal*. In order to understand why, I have to turn to the other main difference between essentialism and my account of modal properties.

Not every essentialist account of properties ascribes causal powers to properties.[22] But they all agree that causal powers determine what *has to* happen in a world. Modal properties then ground primarily necessities. But first, since everything which exists in a world has properties which, in turn, necessitate what happens in that world, it is hard to see how there could be *different* worlds at which the *same* properties are instantiated as are in the actual world. This raises a difficulty for such accounts given that essentialists typically understand 'real necessities' as truth in all possible worlds.[23]

More importantly, essentialism cannot explain how *any* contingency can exist in a world at all. Suppose that at time t_1 the properties in the world are distributed in a way w_1. Since each of these properties has powers which metaphysically necessitate what will happen at t_2, everything which happens *then* is metaphysically necessary as well. And the same is true, of course, about time t_2, when the properties are distributed in a way w_2, and what happens afterwards. The essentialist world seems to be a necessitarian one: only those things can happen in it which actually happen.

It is important to emphasize that the sort of necessity which essentialism entails is not merely the sort of nomic necessity which, according to some incompatibilists, follows from the truth of physical determinism. Incompatibilism is the view that the truth of physical determinism entails that agents never have the ability to do otherwise. However, interestingly enough, the most important argument for incompatibilism, the so-called consequence argument has nothing

---

[22] Ellis, for instance, ascribes causal powers or dispositions to kinds of objects. In this respect, my account of dispositions is similar to Ellis (1999).
[23] See Shoemaker (1998). Ellis is explicit about this, see op. cit., 120, but other necessitarian theories of laws also seem to assume this account of metaphysical necessity.

specifically to do with agents' actions.[24] If the argument is sound, it proves that in a world in which physical laws are (a) deterministic and (b) constitute a closed system (i.e. there are always rules which determine how to 'sum' results when more than one law applies to the evolution of a certain system), nothing else can happen than what actually does.

I have doubts about the soundness of the consequence argument, but this does not matter here.[25] What is important in the present context is that the alleged necessity of all events which is supposed to be established by the consequence argument is at most *hypothetical*, and not absolute and metaphysical. All the consequence argument shows, if sound, is that every event in the world has to happen *given* the contingent laws of physics and the initial conditions of the universe. If the laws were different, other events could happen. But in an essentialist world, laws could not be different since they are metaphysically entailed by the properties and/or the powers. And given that physical laws are time symmetric (or more technically, they are time-reversal invariant), the initial conditions of the universe must also be the same as they actually are. Thus, if essentialism is true, there is no logical place whatsoever for metaphysical contingency in our world.

I do not have any argument that could prove beyond any doubt that we do not live in a necessitarian world. But I do not think either that essentialist metaphysics would give us strong enough reason to accept necessitarianism. If I had to choose, I would certainly reject essentialism rather than accept that everything that happens in the world does so by metaphysical necessity. The essentialist might want to reply that my worry is unjustified since necessitarianism follows only if we assume that every power is deterministic. If there are indeterministic powers and processes in the world as (some of) current physics suggest there are, then of course there is contingency as well.[26]

I do not find this answer satisfactory at all. First, it explains only, if anything, the *physical* conditions in which contingency can exist. It tells us nothing about the metaphysical nature of contingency in a world. In general, it seems to me very implausible to infer from the truth of a *physical* thesis the *metaphysical* impossibility of contingency. But if essentialism is true, then such an inference should be sound. Metaphysical necessitarianism should follow from a possible interpretation of fundamental physical laws.

---

[24] The argument exists in many different versions. Its classical formulation is found in van Inwagen (1975).
[25] I discuss the argument in more detail in my Huoranszki (2011).
[26] This response is explicitly endorsed, for instance, in Heil (2017).

But further, how are we supposed to understand 'indeterministic powers'? There are some accounts of powers which refuse to accept that the ascription of powers entails necessary connections.[27] But the essentialists reject Humanism not only because they believe that properties are modal and not 'qualitative', but also because they do accept necessary connections. Properties in virtue of their powers ground metaphysically necessary laws, which in turn make the occurrence of events which are their consequences necessary too. Now we are told that, in order to find logical place for contingency in the world, (at least some) properties' powers must be indeterminsitic. But then, the relevant properties are not powers at all (except by name), since they only ground certain possibilities, not any necessity.

But third, and most importantly, as I argued in Chapter 3, physical properties do not ground any necessities, irrespective of the truth or falsity of determinism. Most, if not all, dynamical laws are true only *ceteris paribus*. The grounds of such laws are particulars' and substances' properties that *are* dispositions, and not properties which have, or are, powers. The instantiation of these modal properties determines either what things can do or what they tend to do in some typical circumstances. Hence modal properties are not causal powers in the essentialist sense. They are either abilities or dispositions in the sense in which I characterized such properties earlier.

The constituents of actuality

Our investigations have shown that we do not have any clear idea about the content of those allegedly nonmodal properties on which the world-account of contingency is based. Moreover, since worlds should be identified with reference to the distribution of such properties, we could not have any clear idea about the nature of the actual world either. However, many philosophers hold that the alternative account of properties which takes modal properties to be fundamental is exposed to a similar problem. The worry is that the nature of modal properties is bound to be as 'mysterious' as, if I am right, the nature of 'fundamental qualities' are.

However, I do not think that there is anything mysterious about modal properties. And there is nothing mysterious about the nature of nonmodal

---

[27] In both of their books on causal powers, Mumford and Anjum argue against the necessitarian interpretation of powers. See Mumford and Anjum (2011) and Anjum and Mumford (2018).

properties either, if they are properly understood. But, according to what seems to me the proper understanding of their nature, such properties cannot be metaphysically fundamental. The existence of nonmodal and hence non-dynamical properties depends on the existence of modal and hence dynamical properties. This means that my account of the nature of contingency does not entail 'pandispositionalism' either in its weaker or in its stronger sense.

In the latter sense, 'pandispositionalism' is the idea that 'everything is dispositional'. This view is a strange mixture of two metaphysical convictions about the world. The first concerns the rejection of Aristotelian 'primary substance', or more generally, any such ontology which does *not* aim to reduce the category of bearers of properties to properties themselves. Historically, perhaps the most important critics of the idea of substance in this sense were empiricists like Hume and Russell.[28] In Russell's world, particulars are 'compresent' qualitative universals. In Hume's world, they are collections of 'sensible qualities', where such qualities are understood as momentary sensible property-instances ('impressions') the existence of which is metaphysically independent of the object (body or chemical substance) to which they belong. If we follow these empiricist ontologies and hence try to understand objects and matter as 'compresent' collections of properties or momentary property-instances, *but* we are also convinced – contrary to this tradition – that these properties are modal rather than 'qualitative', then strong 'pandispositionalism' straightforwardly follows.[29]

However, if we reject this kind of empiricist metaphysics, as I think we have strong reason to do, then there is no sense in which objects, matter or spacetime should, or in fact could, be understood as 'dispositional'. For what *can* be 'dispositional' – that is to say, modal – are the *properties*, and not the objects, particles and chemical substances which are their bearers; or the places which they occupy. It is true that some objects, which do not actually exist, can exist, but this does not make the category of object modal. In fact, in the account of contingency I am arguing for, the possibility of an object depends on the actuality of the instantiation of some modal properties by some other. For the

---

[28] About Hume's criticism of substance see especially Hume *A Treatise of Human Nature*, Book I, Part IV, sections iii and iv. Russell was sceptical about persisting particularity and the notion of substance throughout his long carrier. For his latest account of bodies and substances as 'complexes of compresent' qualities, see Russell (1948), Chapter VIII.

[29] One might wonder, of course, whether anyone was, or is, *actually* committed to this version of 'pandispositionalism'. The eighteenth-century physicist and mathematician Boscovich is often mentioned as an example on the ground of his claims about the fundamentality of forces. But to decide whether he was really a 'pandispositionalist' would require some more detailed investigation of his views, which I cannot endeavour here.

existence of an object is possible only if there are actually instantiated abilities of some form of matter and some object(s), the exercise of which would result in the (possible) object's coming to be.

More generally, the modal property-based account of contingency that I have defended in this book is incompatible with any 'one category ontology', in which individuals, like ordinary objects, organisms, physical particles and some of the systems constituted by them, or the stuff of which they are constituted (simple or composite chemical substances), should be understood as mere constructs of properties or tropes.[30] Properties and tropes need bearers as much as bearers need properties and tropes. But this mutual ontological dependence is certainly not sufficient reason for trying to reduce one of these fundamental categories to the other.

The actual world, to use Descartes's metaphor, is 'filled in' by bodies and matter. And being an object or a kind of chemical substance means precisely to be a bearer of properties.[31] The concept of bearers of properties can perhaps be acquired only through the exercise of our capacity of abstraction. For we can distinguish the bearers of properties from the properties exemplified by them only in thought. In nature, they always exist together: there is no object, matter and so on without some properties that they have at a moment. But this does not mean that physical particles, objects or matter are merely abstract entities and hence cannot fill in space.

Suppose that what we call *our* concrete universe is the spacetime system in which we live. The view I defend is that what 'fills in' space here are particular bodies and their constituting matter (and perhaps also processes like radiation and transmission of energy if they are distinct from matter), and not properties. Properties themselves do not 'fill in' any space because all properties are abstractions from concretely existing particles, physical bodies and matter. Properties are abstract to varying degree; but being abstract does not mean that they are 'unreal' in any sense. It is maximally specific – that is, the least abstract – properties of objects and matter which determine how they will move, change, act and interact with each other; or which new objects chemical substances

---

[30] For attempts to work out such 'one-category ontologies' based on tropes see Williams (1953) and Campbell (1990).

[31] The 'filling in' metaphor was popularized in current literature by Blackburn (1990), but the expression was used earlier by Descartes's in his Second Meditation to characterize material *bodies* (or 'corporeal substances'). The standard English translation reads as 'occupy space' (see Descartes 1984/1641, 14), but in the original Latin we find 'spatium replere', and the French translation authorized by Descartes also says 'remplir un espace'; that is: filling in space.

can constitute; or how objects begin or cease to exist by the disintegration or reintegration of their constituent matter, and so forth.

Since properties are abstract, properties *themselves* cannot fill in any space. They can 'fill in' space by being instantiated by their bearers. However, their bearers must instantiate some properties, among them some maximally specific modal properties, in order to exist. They cannot exist just by having *a* shape, having *some* mass, having *a* charge, having *a* colour and so on. They must have some specific shape, mass, charge, colour. Similarly, they cannot be just fragile, they must have a specific degree and sort of fragility; they cannot be simply nutritious, they must be able to nourish to certain sort of organisms with certain features and so on. And further, as I argued in detail earlier, most of particular objects' (or samples of substance's) abilities are extrinsic properties sensitive to the changes of their (narrower or broader) environment. Thus, in brief, what 'fills in' our universe's space are *objects and matter having certain maximally specific modal properties at a time*.

'Pandispositionalism', however, can be understood in a weaker sense as well, according to which not everything *simpliciter*, but every property, is (a) modal and (b) *only* modal.[32] The second conjunct is necessary for some metaphysics to be pandispositionalist, because, according to a rather broadly accepted view, (fundamental) properties can be *both* modal and nonmodal.[33] And, although this is partly a matter of convention, I would not qualify this view as pandispositionalist, since it denies that every property is modal *simpliciter*.

The idea that properties are both modal and nonmodal needs to be distinguished from my account of nonmodal properties, according to which such properties *depend* on modal properties, but they are not identical with some modal properties. In fact, as some others,[34] I have difficulty with understanding the motive for, and the meaning of, the claim that some properties are both modal and 'categorical' or 'qualitative' (e.g. nonmodal).

As to the meaning, *prima facie* at least, this claim about identity seems to have a rather different status than some other, better-known claims about 'property-identity' do in contemporary metaphysics. It is not like, for instance, the claim that water is identical with H_2O. Rather, it sounds like saying that 'The round is square' and hence 'The round is non-round'. The ascription of a dynamical

---

[32] Although others already presupposed this interpretation earlier, it was Anjum and Mumford who explicitly define 'pandispositionalism' as a thesis which concerns *only* properties. See Anjum and Mumford (2018, 8).

[33] The first who suggested the 'powerful qualities' account of properties was C. B. Martin; see Martin (1994). Later representatives of this view include, among others, Heil (2003), Strawson (2008) and Williams (2019).

[34] See Taylor (2018) for a very lucid exposition of the problem.

property has, *by its nature*, implications concerning how matter and objects which possess those properties change, act and interact. The ascription of a 'categorical' properties cannot, *by definition*, have any such implications. How can then be these two kinds of properties identical?

But even if we can get somehow around this difficulty, we need to recognize that the very idea of 'property-identity' is far more contentious than it is usually assumed. The most often-invoked examples are far from being clear cases of the alleged identity of properties. Water is certainly not identical with H_2O molecules, since if it were, then ice should also be identical with H_2O molecules; but water, given Leibniz's law, cannot be identical with ice (they differ in respect of countless physical, chemical and sensible features). And we cannot save the claim about identity by adding that 'Water is an ensemble of H_2O molecules in a certain type of kinetic state', since, although this *is* true, it is *not* an identity-statement at all. It expresses a nomological relation between the behaviour of the constituting matter (in this case H_2O molecules) and some macroscopic states. This nomological correlation is certainly grounded in a modal property of the ensemble of H_2O molecules: that they would constitute water if they were in a certain kinetic state. But all this has nothing to do with any kind of 'property-identity'.

Be this as it may, it is also unclear what exactly *motivates* the identity claim about modal and 'categorical' properties. Those who reject first-order, fundamental modal properties often do so on the ground that modal properties cannot explain how anything can be actual. For instance, David Armstrong says that if particulars had only modal properties, bearers of properties would always 're-packing their bags as they change properties, yet never taking a journey from potency to act', since to act 'on this view is no more than a different potency' (Armstrong 1997, 80). My suspicion is that those modal-property realists who insist that modal properties must be 'identical' with 'categorical' properties share Armstrong's view. They accept that we cannot make sense of actuality unless we suppose that fundamental modal properties are somehow also 'categorical'.

But I do not see how worries about actuality follow from the essential modal nature of properties. Such worries seem to be based on the supposition that modal properties or their exercise are just 'mere potentialities'. But this is a confusion. Of course, an *act* cannot be a 'potency' *on any view*. Often – although as we have seen in the previous chapter, not always – when a modal property is exercised, an object changes with respect to some of its properties. What we can consider as an 'act' (or rather as a resultant state or event of the exercise of a modal property) is this actual *change* (or in other cases: non-change, stability or persisting state). And an actual change is, of course, not a 'mere potency'.

It is difficult to understand why a glass, when it gets hardened, or a battery when it gets charged, would *change* merely 'potentially' and not 'in actuality'. Such changes cannot be directly perceived, indeed; but neither can be most other changes with respect to properties that are supposed to be 'categorical' (e.g. changes with respect to the 'microstructure' of macroscopic objects' constituents). And further, sperms and eggs have the ability to produce zygotes, but how are we supposed to understand the claim that when they do so *the zygotes' coming to exist* is a 'mere potency'? A nearby magnet has the ability to divert the position of the pointer of a compass, but what does it mean to say that *the movement of the pointer* is a 'potency'?

As I mentioned already in the first chapter, there is a sense in which material objects, particles or chemical substances must have certain 'potentialities' if they have modal properties at all. But first, modal properties *themselves* are no more 'mere potentialities' than their bearers' actual changes with respect to them are. What might be 'only' potential is the *exercise* of the abilities and dispositions, and hence the occurrence of events or the production of certain objects or chemical substances which would result from their exercise. But neither the *possession* of modal properties nor the events and processes of their *acquisition* need to be 'merely potential'. On the contrary, what constitutes actuality are (*i*) objects', particles', chemical substances' and so on which have modal, dynamical properties, and (*ii*) the objects and events which result from their exercise.

Thus, we need not commit ourselves to some controversial view about the identity of modal and nonmodal properties in order to grant the reality and fundamentality of modal properties. But this does not mean that we need to deny that objects and matter can have properties which are nonmodal either. What we do need to admit is that the nonmodal is always grounded in some modal. The nonmodal, non-dynamical properties cannot be 'fundamental' or ontologically prior to the modal properties of objects. There are nonmodal properties in the world only because there are some modal ones.[35]

How should then we understand nonmodal properties? I suggest that *some of the most abstract properties* that matter, objects, particles and their systems

---

[35] Thus, it is false that objects can have modal properties only because they have nonmodal ones, together with some contingent laws that 'govern' the dynamics of the redistribution of such nonmodal properties in the world. But this is the idea on which all anti-realists or 'categorialists' about modal properties, from J. L. Mackie to David Armstrong and David Lewis, agree, despite their rival views about the nature of properties and laws.

can possess should be interpreted as nonmodal. We have seen in Chapter 5 that, although abilities can be maximally specific, they cannot be maximally generic. Here I want to argue that there is a level of generality and abstraction at which the modality of properties 'fades away' as it were. At that level, the ascription of properties themselves does not entail how objects which instantiate them can act or interact anymore. However, and most importantly, properties at that level of generality can exist and can have instances only if there are things in the world that instantiate some more specific properties from which the nonmodal properties are very generic abstractions.

More precisely, the existence of nonmodal properties depends on the instantiation of some modal properties in the sense that (a) the instantiation of every nonmodal property presupposes that their bearers have maximally specific modal properties which determine how they *can* (or tend to) change or stay stable and hence persist, and how they can (or tend to) act and interact with each other; (b) the so-called nonmodal properties are such highly abstract properties which are abstracted from some logically interdependent modal properties. Expressed otherwise, the central idea here is that, at a sufficiently high level of abstraction, the properties can lose their dynamical character and hence can be understood as being 'categorical'.

To illustrate this idea, consider the example of shape, one of the most often cited examples of allegedly 'categorical' properties. In which sense are shapes nonmodal properties? And on what ground can we ascribe such properties to a specific chunk of matter? It seems that we can ascribe such properties on two kinds of ground: (a) how an object having a certain surface would *look* to us when it is observed in certain conditions or which (series of) tactual impressions we would have if it were touched. (b) how things of its surface would move if certain forces were exercised on them. Hence the ground of ascribing a certain shape is essentially connected to how objects having it would interact with other objects or organisms in some specific circumstances.

For instance, some balls are such that if they were observed from any perspective, they would look round. They are also such that if a particle moved around on its surface, then its distance from the object's centre would always remain the same. And obviously, there are also countless other physical properties which we naturally associate with the objects having the shape of a ball. Moreover, and this is the crucial point, these modal properties are not independent. For *if* an object is such that if a particle moved around on its surface then its distance from the object's centre would remain the same, *then* the object must also be such that it would look round, if it were observed from any perspective. And finally, the interconnections of all these properties can be explained by ascribing the abstract geometrical property of sphericity to the object.

Since 'spheric' is an abstract geometrical property, and abstract geometrical properties *themselves* are not modal, there is a sense in which it is true that the specific chunk of matter we consider has (instantiates) a nonmodal property. But, importantly, the ground of ascribing this property must be the countless interconnected specific modal properties which an object must possess in order to instantiate the abstract geometrical one. It seems a rather counterintuitive thought that an abstract geometrical property, which shape in fact is, can, in itself, contribute to the behaviour and interactions to some objects. In contrast, it seems very intuitive that the ground of ascribing such properties are the *countless different but interdependent modal properties*, specifiable with the help of conditionals, which we naturally associate with the abstract geometrical property of a given object.

Shape, of course, is just one possible example of such properties. Other often mentioned examples are spatial locations and distance, magnitudes, numerosity, structure or structural composition.[36] However, it is interesting to note about all these examples that each of them concerns some rather abstract, typically spatial and geometrical, properties. Hence it might not be misleading to call them the essentially *Cartesian properties of matter*. Some of them (shape, size, location, structure) are clearly geometrical. Others (magnitudes, numerosity, structural composition) are arithmetic, but in the broad Cartesian sense in which geometrical-structural features can be arithmetically represented (e.g. by using coordinates and other mathematical techniques to capture dimensions). Thus, the aforementioned argument applies to them just as much as it applies to shapes. Such properties are abstractions from an interconnected net of modal properties, which are less generic, non-Cartesian properties of matter.

This, in fact, also explains why such properties, unlike modal properties, can be attributed to objects and particles – as well as to the systems composed by them or to the chemical substances which constitutes them – without having any implications about how their bearers change, act or interact. For abstract geometrical properties are simply not the kind of properties that in themselves determine any action. Properties that do have implications

---

[36] See, for instance, Molnar (2003), Chapter 10; Bird (2016, 355–6); or Ellis (2010).

as to how their bearers behave and interact are always some of those more determinate properties the instantiation of which justifies the ascription of the most abstract nonmodal properties. But those other properties are, of course, all modal properties.

Cartesian properties of matter are perhaps the clearest examples of nonmodal properties. There is a sense though in which we can also understand natural kinds and some 'fundamental' physical properties as nonmodal as well. They are obviously less abstract than the geometrical properties discussed so far. However, they are more abstract than many ordinary abilities and dispositions. Natural kinds and some physical properties have a special ontological role because they are essentially classificatory properties. Classification itself is abstract enough for not having direct implications to objects' behaviour. But classification is directly metaphysically linked to the ascription of modal properties that are dispositions.

As we have seen earlier in Chapter 4, we often need natural kinds to explain the link between dispositions and conditionals, though we do not need them in order to explain abilities. This is not an accident. Many dispositions ground nomological tendencies, and some sort of features or behaviour can be tendentious in this sense only if the objects to which we ascribe the relevant dispositions belong to one specific species or have certain physical properties, but exceptional or hence non-tendentious if they belong to another. For this reason, natural kinds and certain physical property-types play a crucial role in our account of dispositions.

However, for exactly the same reason, it is obvious that the ascription of natural kinds depends on the ascription of some modal properties. Natural kinds (and the relevant physical-type properties), like Cartesian properties of matter, are abstracted from interdependent modal properties. Natural kinds are less abstract than Cartesian properties, and they have a distinct explanatory role: they explain why certain logically unconnected dispositional properties are non-accidentally co-instantiated by the same individuals. Hence although membership in a natural kind itself has no direct implication about how matter, objects or organisms will behave, they can have such implications indirectly through their connection to the associated nomic dispositions. The sense then in which natural and physical kinds are nonmodal is that they have no direct implications concerning the possible or regular behaviour of their instances. But they do have direct logical connections to the nomic and behavioural dispositions which their members possess.

Idealism

So far, I have argued that nonmodal or 'categorical' properties are the most abstract kind of properties the existence of which depends on concrete objects' more specific modal properties. Some philosophers hold, however, that the ontologically fundamental nonmodal properties are *sensibly occurrent qualities* which do not seem to be abstract. Just to the contrary, it might seem that there are not any properties that can be more concrete than such qualities are.

These philosophers then, following Hume, argue that the existence of modal properties *must* depend on the occurrence of such qualities. But if (*i*) every property of material and physical objects is modal and (*ii*) the instantiation of any modal properties depends on the instantiation of some nonmodal property, which (*iii*) are all sensibly occurring and hence mind-dependent, then the structure of material reality cannot be independent of the mind. Consequently, the non-fundamentality or derivativeness of modal properties must imply a form of idealism.[37]

Now, I certainly agree with the first premise of this argument. Since material properties ground possibilities and laws, and thereby ground the ways in which objects and substances can, or tend to, change and interact, these properties must be modal. However, as I argued above, I do not think that the existence of modal properties must depend on the nonmodal. Nor does it seem to me true that the sensibly occurring *properties* are nonmodal. Surely, sensible occurrences are not modal because sensible occurrences are not properties but events. But this does not mean that the properties which such events manifest are not modal.

As we have seen in the previous chapter, the modal property-based account of contingency assigns a special role to events. According to that theory, contingency is an essentially *dynamical* phenomenon. Persisting entities are contingent in the sense that they can come about and pass away and hence their existence depends on their 'birth' and 'death', which are special sorts of events. Chemical substances which constitute matter and objects can also come about and pass away by fusion and decomposition. The existence of contingent particulars, objects and persons depends on modal properties the exercise of which results in their coming to be and in their persisting during a certain period of time.

However, all this does not mean that the modal property-based account must assign a general *ontological primacy* to events. For, as I also argued in the previous

---

[37] This is how Howard Robinson argues from the primacy of the qualitative to idealism in Robinson (1982, 114–15).

chapter, what happens or exists contingently depends on the nature of those modal properties that we ascribe to some sample of chemical substance, to some object, organism or person, or to some of their ensembles. What can and cannot *happen* depends on what objects, persons or substances can or cannot *do*: some cups can break; therefore, there must be a sense in which it is possible that they break even if, actually, they never do; some people can win an election; therefore, there must be a sense in which it is possible for them to win an election, even if, actually, they lose. Contingent possibilities are grounded in modal properties instantiated by objects and substances in spacetime.

There is no reason to suppose that sensibly occuring qualitative properties are different in this respect. Such properties must be properties *of* something: either the properties of objects and matter perceived or the properties of the subject perceiving them. Hence, what they are must depend either on the modal properties of objects perceived or on the modal properties of the perceiving subject. Actuality can be interpreted as the distribution of sensibly occurrent qualities only if we have already committed ourselves to a specific form of idealist metaphysics, in which not only modal properties, but also material objects, substances as well as the perceiving subjects are taken to be derivative and hence dependent entities.

If, however, we reject this form of idealism, then there is no reason to assume that sensibly occurring properties are nonmodal. For such properties obviously ground objects' interactions with those creatures who have the proper sensitivities; that is, the abilities to mentally respond to the presence of such properties in a specific manner in some appropriately defined conditions. After all, this is what makes them *sensibly* occurring qualities. How we understand those events must depend on which modal properties are exercised. The fact that I cannot hear voices with my eyes and see colours with my ears is explained, primarily and fundamentally, by the fact that hearing voices are the exercise of my auditory capacities and seeing colours is an exercise of my visual capacities.

Hume's philosophy and, more generally, the phenomenalist tradition just assumed that sensibly occurring qualities, and only such qualities, reveal their own nature without having implications as to what else exists, or can occur, in the world. But the idea that the qualities which occur reveal their own essence is far from being an obvious truth. According to a certainly no less plausible account, sensibly occurrent qualities are the results of the exercise of two abilities: a macroscopic physical object's ability to affect a subject in a certain way and the subject's (physical or mental) ability to respond to that affection in a certain manner. There are no sensible occurrences without the exercise of those abilities and hence what they are or can be depends essentially on the properties of objects (some of whom are also *subjects* of experiences) as any other events in the world do.

When *I* look at a plant's leaves and *they* look green to me, the *event* of my seeing something is directly epistemically accessible to me. But this does not mean that that event directly reveals the very nature or essence of any *properties*. In fact, how much one thinks is revealed about the real nature of colour properties by that event depends on the theory of colours that one accepts. Perhaps colours are essentially wavelengths of electromagnetic radiations; or microphysically characterizable features of the surfaces of some opaque object or substance.

I do not have a view of what colours, or sensible qualities in general, 'really are'. But to say that the nature of sensed properties must be revealed by their being sensed seems to me rather implausible. The nature of (the act of) our *sensing* certain properties in the world might be revealed to us, but this should not be confused with the nature and essence of the very property sensed. When we touch an object with a hand, we can feel its texture, temperature and the shape of its surface. These are its distinct properties, but their essence or nature is certainly not revealed merely by our ability to sense them when touching it. Moreover, the very act of sensing them is obviously the manifestation of *our* abilities as well: the abilities that we are exercising when we see, hear, touch, etc. something.

The system of beings

Having said what I think about the nature of so-called nonmodal, 'categorical' or qualitative properties I hope I can be relatively brief in answering the standard objection to the view that first-order modal properties are ontologically fundamental. The objection claims that granting the fundamentality of first-order modal properties entails a problematic regress, called 'the regress of pure powers'.

This objection to the fundamentality of modal properties assumes, first, that the nature or identity of these properties must depend on what would manifest them. Second, it claims that if modal properties were fundamental, then a modal property could be manifested only by the occurrence of some other modal properties. And then, it follows that, unless we grant that modal properties must depend on the nonmodal, we end up in a regress, the 'regress of pure powers'. Finally, this regress is assumed to be vicious, because it entails that in order for a modal property to have a determinate nature, an infinite series of dependent modal properties needs to be postulated.[38]

The first problem with this argument is that it is difficult to understand how a property could have a nature *without* its instantiation having implications with regard to other things in the world, since 'nature' itself, in this sense, seems to be a modal notion. But further, even if the existence of modal properties must depend on something else – a requirement which realists about modal properties can of course reject – the alleged dependence of modal properties on other modal properties does not entail that the series of dependence must be infinite. Rather, the dependence is *mutual*, even if it is not always direct.

The exercise of modal property A might entail that an object changes with respect another modal property B, which, when exercised, involves a change with respect to modal property Δ, and then it involves a change with respect to property Γ … and so until Ω. But then, the exercise of modal property Ω might entail that an object changes with respect to B, let us say, in circumstances that are partly characterized by Γ, and so forth again.

Modal properties are then taken to be *interdependent* as, certainly, any plausible account of modal properties must admit (see especially the arguments in Chapter 5). There is not any threat of 'vicious infinite regress' here.[39] In fact, it seems to me entirely inconceivable how one could give *any account* of properties, modal or not, which does not assume such interdependence. For how can one give an *account* of anything without reference to some (other) properties?

The objection then might be that, instead of being threatened by a vicious regress, realism about the first-order modal properties entails a 'vicious circle'. Circle indeed, by why should it be 'vicious'? Can we indeed understand *anything* without some such circles? Does it make sense to assume that real properties must exist, like Cartesian substances, independent of anything else and not only as parts of *a system* of properties? On the contrary, it seems to me that properties, by their very nature, can exist *only* as a system.[40]

---

[38] For the clearest early statement of the argument see again Robinson (1982, 114–15); the argument has been recently defended by Ingthorsson (2015). As Bird, who also gives a fair summary of the different versions of the arguments, correctly notes though, the argument rest on the presupposition that modal properties *must* depend on some other properties while qualities do not. And this is exactly the supposition which modal realists reject. See Bird (2007b).

[39] Similar arguments have been presented by Holton (1999) and Bird (2007). For criticism see Lowe (2010) as well as Heil (2003, 2010). Both Holton and Bird rely, however, on an analogy with graph theory. I do not find that analogy enlightening, and it is potentially misleading. For the interdependence of the *content* of properties is rather dissimilar to the formal relations definable among contentless nodes of a graph.

[40] I take this to be one of Hegel's deepest insight and contribution to the history of the metaphysics of properties.

First, observe that the same circularity argument that is raised against the existence of first-order modal properties can apply to almost any account of the scientifically discoverable 'theoretical properties'. We will never understand the 'nature' of mass without the concept of 'force', solubility without being a solvent, radiation without absorption and so on. Theoretical properties are properties that figure in laws which provide a link between (or rather: among) such properties.[41] Without such a link these properties could not have any content or 'determinate nature'. Thus, the interdependence of properties as properties seems to be the rule, rather than an exception.

Perhaps those who insist that such interdependence entails a vicious circle want to respond that all this shows is only that scientifically discoverable properties cannot have a 'determinate nature' *unless* they are also connected to sensibly occurring qualities. But this would mean that the material world, that is the world characterized by material and physical properties, cannot exist without being sensible creatures in the world. This again entails a form of idealism. But we are certainly not bound to accept idealism on this ground. It is one thing to grant that, without having and exercising our capacities of sensation and perception, we would lack the sort of epistemic access to the material world which we actually have. And it is quite another to infer from this that the *existence* of properties must depend on the possibility of our exercising these capacities.

More generally, the existence of *any* property whatsoever requires sameness and difference in some respect or other. But in *which* respect exactly can only be explained with reference to some further properties, which must also involve sameness and differences in some further respect. Thus, it does not seem to be logically possible to identify the nature of any property whatsoever without having a system of properties of which it is a part. Every property must be *distinct* from some other in respect of the *same* determinable; and further, the sameness of a determinable presupposes the distinctness from other determinables. As Hegel saw it clearly, there are no qualities in the world without being a system of such qualities, and we can hardly understand the nature of such qualities without their being a part of a system.[42]

It is time then to finally conclude and sum up my views of the problem of actuality and the nature of contingency in general. In my view, the best account

---

[41] The most popular logical analysis of theoretical terms, which relies on the formal device of the so-called Ramsey-sentences, is obviously holistic and exploits the interdependence of the content of the respective properties defined. See, among many others, Mellor (1995, 191–6) and Lewis (2009).

[42] See especially Hegel's brilliant analysis of how qualitative distinctions in the world are derivable from the very idea of existence at the beginning of *The Science of Logic*, Book One, Section 1.

of contingency is not compatible with any form of 'one category' ontology, which takes the actual world to be the totality of tropes or properties. The world is not only a distribution of 'qualitative', or any other sort of, properties. Rather, the world is the totality of objects and substances (in the broadly 'chemical' sense of substances as forms of matter that may or may not constitute some macroscopic objects or specific types of physical systems) which exist in one spacetime. It is one thing to say that there are no objects or substances *without* instantiating some properties. It is another that the category of material substances and objects in spacetime can be 'reduced to' the distribution of 'categorical' or sensibly occurring qualitative properties.

Further, modal properties play a fundamental role not only in our understanding of the nature of contingent possibilities but in an account of the structure of actuality as well. Some modal properties are actually exercised sometimes, and the resulting states of their exercise explain the dynamical structure of our ever-changing world. Material objects and systems are formed, and persist for a while, when dynamical-modal properties are actually exercised, and hence possible events are actualized. And the nature of processes as well as the nature of events and states in which they result can be best understood only with reference to those modal properties that they manifest.

Some modal properties manifest themselves by activating our senses. But this does not imply that we can understand their nature simply by sensing certain 'qualities' (if it is really that which we sense). Modal properties can be specified with reference to their contribution to what happens or exists in the world only because we understand what happens in the world *intellectually and conceptually* as much as sensibly. To this extent, John Mackie was right when he called the modal realist position 'rationalist'.[43]

That abilities, and modal properties in general, are metaphysically prior to the events which are their exercise does not contradict the intuitive thesis of the priority of the actual. For what is 'merely' possible is the *exercise* of an ability and hence the occurrence of the results of its exercise. But the *facts* about abilities as first-order modal properties being instantiated at a certain spacetime point or region by a certain object or substance are not 'mere potentialities'. Such facts constitute actuality.

Thus, Aristotle's thesis about the 'priority of the actual' ought to be interpreted as a thesis about the bearers of properties (objects, kinds of matter, systems of

---

[43] Even if, as he himself emphasized, we should be careful to distinguish the epistemic problem from the ontological one; see Mackie (1977, 362).

objects in spacetime), without which nothing could exist in the world. It is this interpretation of the priority of the actual, or priority 'in substance', that I find plausible. In contrast, I find it rather implausible that things that can *happen* are metaphysically prior to things that are bearers of modal properties. There are no 'free-floating changes' that are not the results of the exercise of some abilities; which in turn are borne by some substance or object. What we call 'qualitative change' only helps *us* to (partially) specify certain modal properties; such changes need not reveal the real nature or essence of those changes.[44]

The idea that modal properties cannot be first-order properties, which is to say that they must 'depend' on some other, 'categorical' properties, or that their nature is determined by what is 'qualitative', is a legacy of phenomenalism. Phenomenalism is a radically impoverished ontology. In phenomenalism, objects, forms of matter and their properties are understood in terms of 'sensible qualities'; or, rather, in terms of the qualities of (the act of) our sensing them. But this is not how we commonsensically or scientifically perceive the world. We naturally perceive the world as filled with matter and populated by objects which have modal properties; that is, with such properties that ground objects' behaviour and elicit some of our sensations.

In general, phenomenalism radically underestimates the importance of our intellectual capacities in the natural understanding of the world. Perceiving the world is not the same as having mere sensations. Perceiving the world helps us to understand it, and understanding the world is also a way to interpret it. The more sophisticated our understanding is, the more we conceive the world conceptually rather than merely sensually. And the most important way to understand the world conceptually is to describe it as a world with objects with abilities and dispositions, not as a collection of sensibly occurring qualities. For, we understand the structure of the world intellectually when we understand what things in it can, or tend to, do.

---

[44] It might be argued that qualities can exist without any *object* like the blueness of the sky or the smell of the fresh air allegedly do. However, they cannot exist without some matter in the sense of chemical substances or some more or less well-identifiable system of particulars. The void does not have any colour; neither does it smell.

References

Adams, R. M. (1994), *Leibniz. Determinist, Theist, Idealist*, Oxford: Oxford University Press.
Alvarez, M. (2017), 'Are Character Traits Dispositions?', *Philosophy*, 80: 69–86.
Alvarez, M. and J. Hyman (1998), 'Agents and their Actions', *Philosophy*, 73: 219–45.
Anjum, R. L. and S. Mumford (2018), *What Tends to Be. The Philosophy of Dispositional Modality*, Abingdon: Routledge.
Anscombe, G. M. E. (1993/1971), 'Causality and Determination', in E. Sosa and M. Tooley (ed.), *Causation*, 88–104, Oxford: Oxford University Press.
Aristotle. (1984a), 'Physics', in J. Barnes (ed.), *The Complete Works of Aristotle. Volume One*, 315–446, Princeton: Princeton University Press.
Aristotle. (1984b), 'Prior Analytics', in J. Barnes (ed.), *The Complete Works of Aristotle. Volume One*, 39–113, Princeton: Princeton University Press.
Aristotle. (1984c), 'Metaphysics', in J. Barnes (ed.), *The Complete Works of Aristotle. Volume Two*, 1552–728, Princeton: Princeton University Press.
Armstrong, D. M. (1968), *A Materialist Theory of Mind*, London: Routledge.
Armstrong, D. M. (1973), *Belief, Truths, and Knowledge*, Cambridge: Cambridge University Press.
Armstrong, D. M. (1983), *What Is a Law of Nature?*, Cambridge: Cambridge University Press.
Armstrong, D. M. (1989), *A Combinatorial Theory of Possibility*, Cambridge: Cambridge University Press.
Armstrong, D. M. (1997), *A World of States of Affairs*, Cambridge: Cambridge University Press.
Armstrong, D. M., C. B. Martin and U. T. Place (1996), *Dispositions: A Debate*, ed. T. Crane, London: Routledge.
Ayer, A. J. (1936), *Language Truth, and Logic*, London: Gollancz.
Beere, J. (2009), *Doing and Being. An Interpretation of Aristotle's Metaphysics Theta*, Oxford: Oxford University Press.
Bennett, J. (2003), *A Philosopher's Guide to Conditionals*, Oxford: Oxford University Press.
Bigelow, J., B. Ellis and C. Lierse (1992), 'The World as One of a Kind: Natural Necessity and Laws of Nature', *British Journal for the Philosophy of Science*, 43: 371–88.
Bird, A. (1998), 'Dispositions and Antidotes', *The Philosophical Quarterly*, 48: 227–34.
Bird, A. (2005), 'The Dispositionalist Conception of Laws', *Foundations of Science*, 10: 353–70.

Bird, A. (2007a), *Nature's Metaphysics: Laws and Properties*, Oxford: Oxford University Press.
Bird, A. (2007b), 'The Regress of Pure Powers', *Philosophical Quarterly*, 57: 513–34.
Bird, A. (2016), 'Overpowering: How the Powers Ontology has Overreached Itself', *Mind*, 125: 341–83.
Black, R. (2000), 'Against Quidditism', *Australasian Journal of Philosophy*, 78: 87–104.
Blackburn, S. (1990), 'Filling in Space', *Analysis*, 50: 62–5.
Broad, C. D. (1925), *The Mind and Its Place in Nature*, London: Routledge and Kegan Paul.
Butler, D. (1988), 'Character Traits in Explanation', *Philosophy and Phenomenological Research*, 49: 215–38.
Campbell, K. (1990), *Abstract Particulars*, Oxford: Basil Blackwell.
Carnap, R. (1936–37), 'Testability and Meaning', *Philosophy of Science*, 3: 419–71 and 4: 1–40.
Cartwright, N. (1989), *Nature's Capacities and their Measurement*, New York: Oxford University Press.
Cartwright, N. (1999), *The Dappled World: A Study of the Boundaries of Science*, Cambridge: Cambridge University Press.
Charles, D. (1984), *Aristotle's Philosophy of Action*, London: Duckworth.
Choi, S. (2003), 'Improving Bird's Antidotes', *Australasian Journal of Philosophy*, 81: 573–80.
Choi, S. (2006), 'The Simple vs. Reformed Conditional Analysis of Dispositions', *Synthese*, 148: 369–79.
Choi, S. (2008), 'Dispositional Properties and Counterfactual Conditionals', *Mind*, 117: 795–841.
Clarke, R. (2008), 'Intrinsic Finks', *The Philosophical Quarterly*, 58: 512–18.
Clarke, R. (2009), 'Dispositions, Abilities to Act, and Free Will: The New Dispositionalism', *Mind*, 118: 323–51.
Davidson, D. (1969), 'The Individuation of Events', reprinted in *Essays on Actions and Events*, 1980, 163–80, Oxford: Oxford University Press.
Davidson, D. (1971), 'Agency', reprinted in *Essays on Actions and Events*, 1980, 43–61, Oxford: Oxford University Press.
Demarest, H. (2017), 'Powerful Properties, Powerless Laws', in J. Jacobs (ed.), *Causal Powers*, 38–54, Oxford: Oxford University Press.
Descartes, R. (1984/1641), 'Meditations on First Philosophy', in J. Cottingham, R. Stoothoff and D. Murdoch (trans.), *The Philosophical Writings of Descartes Volume II*, 1–62, Cambridge: Cambridge University Press.
Dipert, R. (1997), 'The Mathematical Structure of the World: The World as Graph', *Journal of Philosophy*, 94: 329–58.
Eagle, A. (2011), 'Deterministic Chance', *Nous*, 45: 269–99.
Ellis, B. (1999), *Scientific Essentialism*, Cambridge: Cambridge University Press.

Ellis, B. (2002), *The Philosophy of Nature: A Guide to the New Essentialism*, Montreal: McGill-Queen's University Press.

Ellis, B. (2010), 'Causal Powers and Categorical Properties', in A. Marmodoro (ed.), *The Metaphysics of Powers: Their Grounding and Their Manifestations*, 133–42, London: Routledge.

Ellis, B. and C. Lierse (1994), 'Dispositional Essentialism', *Australasian Journal of Philosophy*, 72: 27–45.

Fara, M. (2005), 'Dispositions and Habituals', *Noûs*, 39: 43–82.

Fara, M. (2008), 'Masked Abilities and Compatibilism', *Mind*, 117: 843–65.

Fine, K. (1994), 'Essence and Modality: The Second Philosophical Perspectives Lecture', *Philosophical Perspectives*, 8: 1–16.

Fine, K. (2002), 'The Varieties of Necessity', in T. Gendler and J. Hawthorne (eds), *Conceivability and Possibility*, 253–81, Oxford: Oxford University Press.

Funkhouser, E. (2006), 'The Determinable-Determinate Relation', *Nous*, 40: 548–69.

Gendler, T. and J. Hawthorne, eds (2002), *Conceivability and Possibility*, Oxford: Oxford University Press.

Glynn, L. (2010), 'Deterministic Chance', *British Journal for the Philosophy of Science*, 61: 51–80.

Glynn, L. (2012), 'Book Review. Getting Causes from Powers, by Mumford, S. and Anjum, R. L. (Oxford: Oxford University Press, 2011)', *Mind*, 121: 1099–106.

Gnassounou, B. (2007), 'Conditional Possibility', in M. Kistler and B. Gnassounou (eds), *Dispositions and Causal Powers*, 151–60, Aldershot: Ashgate.

Goldman, A. (1970), *A Theory of Human Action*, Englewood Cliff: Prentice Hall.

Goodman, N. (1954), *Fact, Fiction and Forecast*, Cambridge, MA: Harvard University Press.

Griffin, M. V. (2013), *Leibniz, Good and Necessity*, Cambridge: Cambridge University Press.

Handfield, T., ed. (2009a), *Dispositions and Causes*, Oxford: Oxford University Press.

Handfield, T. (2009b), 'The Metaphysics of Dispositions and Causes', in T. Handfield (ed.), *Dispositions and Causes*, 1–31, Oxford: Oxford University Press.

Handfield, T. (2010), 'Dispositions, Manifestations, and Causal Structure', in Anna Marmodoro (ed.), *The Metaphysics of Powers: Their Grounding and Their Manifestations*, 106–32, New York: Routledge.

Harré, R. (1970), 'Powers', *British Journal for the Philosophy of Science*, 21: 81–101.

Harré, R. and E. H. Madden (1975), *Causal Powers: A Theory of Natural Necessity*, Oxford: Basil Blackwell.

Hawthorne, J. (2001), 'Causal Structuralism', *Philosophical Perspectives*, 15: 361–78.

Hegel, G. W. F. (1812/2010), *The Science of Logic*, trans. and ed. George di Giovanni, Cambridge: Cambridge University Press.

Heil, J. (2003), *From an Ontological Point of View*, Oxford: Clarendon Press.

Heil, J. (2010), 'Powerful Qualities', in Anna Marmadoro (ed.), *The Metaphysics of Powers—Their Grounding and their Manifestations*, 58–72, New York: Routledge.

Heil, J. (2017), 'Real Modalities', in J. D. Jacobs (ed.), *Causal Powers*, 90–104, Oxford: Oxford University Press.

Hempel, C. (1965), *Aspects of Scientific Explanation and Other Essays in the Philosophy of Science*, New York: Free Press.

Hendry, R. and D. Rowbottom (2009), 'Dispositional Essentialism and the Necessity of Laws', *Analysis*, 69: 668–77.

Hintikka, J. (1973), *Time and Necessity: Studies in Aristotle's Theory of Modality*, Oxford: Clarendon Press.

Holton, R. (1999), 'Dispositions all the Way Round', *Analysis*, 59: 9–14.

Horgan, T. (1993), 'From Supervenience to Superdupervenience: Meeting the Demands of a Material World', *Mind*, 102: 555–85.

Hoefer, C. (2007), 'The Third Way on Objective Probability: A Sceptic's Guide to Objective Chance', *Mind*, 116: 549–96.

Hume, D. (1975(T)), *A Treatise of Human Nature*, ed. L. A. Selby-Bigge, 2nd edn., revised by P. H. Nidditch, Oxford: Clarendon Press.

Hume, D. (1975(E)), *Enquiries concerning Human Understanding and concerning the Principles of Morals*, ed. L. A. Selby-Bigge, 3rd edn., revised by P. H. Nidditch, Oxford: Clarendon Press.

Humberstone, I. L. (1996), 'Intrinsic/Extrinsic', *Synthese*, 108: 205–67.

Huoranszki, F. (2011), *Freedom of the Will. A Conditional Analysis*, New York: Routledge.

Huoranszki, F. (2012), 'Powers, Dispositions, and Counterfactual Conditionals', *Hungarian Philosophical Review*, 56 (4): 33–53.

Huoranszki, F. (2017), 'Alternative Possibilities and Causal Overdetermination', *Disputatio*, 45: 193–217.

Huoranszki, F. (2019), 'The Contingency of Physical Laws', *Principia: An International Journal of Epistemology*, 23: 487–502.

Hüttemann, A. (1998), 'Laws and Dispositions', *Philosophy of Science*, 65: 121–35.

Hüttemann, A. (2021), 'The Return of Causal Powers?', in H. Lagerlund, B. Hill and S. Psillos (eds), *Reconsidering Causal Powers: Historical and Conceptual Perspectives*, 168–85, Oxford: Oxford University Press.

Ingthorsson, R. (2015), 'The Regress of Pure Powers', *The European Journal of Philosophy*, 23: 529–41.

Jacobs, J. D. (2010), 'A Powers Theory of Modality – or, How I Learned to Stop Worrying and Reject Possible Worlds', *Philosophical Studies*, 151: 227–48.

Jacobs, J. D., ed. (2017), *Causal Powers*, Oxford: Oxford University Press.

Johnston, M. (1992), 'How to Speak of the Colors', *Philosophical Studies*, 68: 221–63.

Jubien, M. (2009), *Possibility*, Oxford: Oxford University Press.

Kant, I. (1929/1781), *Critique of Pure Reason*, trans. Norman Kemp Smith. London: Macmillan and Co.

Kim, J. (1993/1976), 'Events as Property Exemplifications', reprinted in his *Supervenience and Mind*, 33–52, Cambridge: Cambridge University Press.

Kim, J. (1993/1982), 'Psychophysical Supervenience', reprinted in his *Supervenience and Mind*, 175–93, Cambridge: Cambridge University Press.

Kistler, M. and B. Gnassounou, eds (2007), *Dispositions and Causal Powers*, Aldershot: Ashgate.

Koslicki, K. (2015), 'The Coarse-Grainedness of Grounding', *Oxford Studies in Metaphysics*, 9: 306–44.

Kosman, L. A. (1969), 'Aristotle's Definition of Motion', *Phronesis*, 14: 40–62.

Kripke, S. (1972), 'Naming and Necessity', in D. Davidson and G. Harman (eds), *Semantics of Natural Language*, Dordrecht: Reidel, 253–355.

Kroll, N. (2017), 'Teleological Dispositions', in K. Bennett and D. W. Zimmerman (eds), *Oxford Studies in Metaphysics*, vol. 10, 3–37, Oxford: Oxford University Press.

Lagerlund, H., B. Hill and S. Psillos, eds (2021), *Reconsidering Causal Powers: Historical and Conceptual Perspectives*, Oxford: Oxford University Press.

Langton, R. and D. Lewis (1999/1998), 'Defining "Intrinsic"', in D. Lewis (ed.), *Papers in Metaphysics and Epistemology*, 116–32, Oxford: Oxford University Press.

Leibniz, G. W. (1989), *Philosophical Essays*, trans. R. Ariew and D. Garber, Indianapolis: Hackett Publishing.

Leibniz, G. W. (1686/1989), 'Discourse on Metaphysics', in D. Garber and R. Ariew (trans.), *Philosophical Essays*, 35–68, Indianapolis: Hackett Publishing Company.

Leibniz, G. W. (1765/1996), *New Essays in Human Understanding*, trans. and eds P. Remnant and J. Bennett, Cambridge: Cambridge University Press.

Levey, S. (2005), 'On Precise Shapes and the Corporeal World', in J. A. Cover and D. Rutherford (eds), *Leibniz: Nature and Freedom*, 69–95, New York: Oxford University Press.

Lewis, D. (1973), *Counterfactuals*, Cambridge, MA: Harvard University Press.

Lewis, D. (1999/1983), 'New Work for a Theory of Universals', reprinted in *Papers in Metaphysics and Epistemology*, 8–55, New York: Cambridge University Press.

Lewis, D. (1986a), *On the Plurality of Worlds*, Oxford: Basil Blackwell.

Lewis, D. (1986b), 'Events', in *Philosophical Papers. Volume II*, 241–69, New York: Oxford University Press.

Lewis, D. (1986/1976), 'The Paradoxes of Time Travel', reprinted in *Philosophical Papers. Volume II*, 67–80, New York: Oxford University Press.

Lewis, D. (1986/1979), 'Counterfactual Dependence and Time's Arrow', reprinted in *Philosophical Papers. Volume II*, 32–52, New York: Oxford University Press.

Lewis, D. (1999/1997), 'Finkish Dispositions', reprinted in *Papers in Metaphysics and Epistemology*, 133–51, New York: Cambridge University Press.

Lewis, D. (2009), 'Ramseyan Humility', in D. B. Mitchell and R. Nola (eds), *Conceptual Analysis and Philosophical Naturalism*, 203–22, Cambridge, MA: MIT Press.

Locke, J. (1689/1975), *An Essay Concerning Human Understanding*, ed. P. H. Nidditch, Oxford: Clarendon Press.

Loewer, B. (2007), 'Laws and Properties', *Philosophical Topics*, 35: 313–28.

Lombard, L. (1986), *Events: A Metaphysical Study*, London: Routledge and Kegan Paul.

Loux, M. J., ed. (1979), *The Possible and the Actual*, Ithaca: Cornell University Press.

Lowe, E. J. (2006), *The Four Category Ontology: A Metaphysical Foundation for Natural Sciences*, Oxford: Clarendon Press.

Lowe, E. J. (2009), *Personal Agency*, Oxford: Oxford University Press.

Lowe, E. J. (2010), 'On the Individuation of Powers', in A. Marmodoro (ed.), *The Metaphysics of Powers: Their Grounding and Their Manifestations*, 8–26, New York: Routledge.

Mackie, J. L. (1973), *Truth, Probability and Paradox*, Oxford: Oxford University Press.

Mackie, J. L. (1974), *The Cement of the Universe*. Oxford: Clarendon Press.

Mackie, J. L. (1976), *Problems from Locke*. Oxford: Clarendon Press.

Mackie, J. L. (1977), 'Dispositions, Grounds and Causes', *Synthese*, 34: 361–70.

Manley, D. and R. Wasserman (2007), 'A Gradable Approach to Dispositions', *The Philosophical Quarterly*, 57: 68–75.

Manley, D. and R. Wasserman (2008), 'On Linking Dispositions and Conditionals', *Mind*, 117: 59–84.

Marmodoro, A. (2017), 'Aristotelian Powers at Work', in J. D. Jacobs (ed.), *Causal Powers*, 57–76, Oxford: Oxford University Press.

Marmodoro, A. (2010a), 'Do Powers Need Powers to Make them Powerful? From Pandispositionalism to Aristotle', in Anna Marmadoro (ed.), *The Metaphysics of Powers—Their Grounding and their Manifestations*, 27–40, New York: Routledge.

Marmodoro, A., ed. (2010b), *The Metaphysics of Powers: Their Grounding and Their Manifestations*, New York: Routledge.

Martin, C. B. (1993), 'Power for Realists', in J. Bacon, K. Campbell and L. Reinhardt (eds), *Ontology, Causality, and Mind*, 175–94, Cambridge: Cambridge University Press.

Martin, C. B. (1994), 'Dispositions and Conditionals', *The Philosophical Quarterly*, 44: 1–8.

Martin, C. B. (2008), *The Mind in Nature*, Oxford: Oxford University Press.

Martin, B. and K. Pfeifer (1986), 'Intentionality and the Non-Psychological', *Philosophy and Phenomenological Research*, 46: 531–54.

Marshall, D. and B. Weatherson (2018), 'Intrinsic vs. Extrinsic Properties', in Edward N. Zalta (ed.), *The Stanford Encyclopedia of Philosophy*, Spring 2018 edn. Available online: https://plato.stanford.edu/archives/spr2018/entries/intrinsic-extrinsic/.

Mates, B. (1986), *The Philosophy of Leibniz. Metaphysics and Language*, New York: Oxford University Press.

Mcdonald, C. (2005), *Varieties of Things. Foundations of Contemporary Metaphysics*, Malden: Blackwell.

McKitrick, J. (2003), 'A Case for Extrinsic Dispositions', *Australasian Journal of Philosophy*, 81: 155–74.

McKitrick, J. (2009), 'Dispositions, Causes, and Reduction', in T. Handfield (ed.), *Dispositions and Causes*, 31–64, Oxford: Oxford University Press.

McKitrick, J. (2010), 'Manifestations as Effects' in A. Marmodoro (ed.), *The Metaphysics of Powers: Their Grounding and Their Manifestations*, 73–83, New York: Routledge.

McKitrick, J. (2018), *Dispositional Pluralism*, New York: Oxford University Press.
Mellor, D. H. (1971), *The Matter of Chance*, Cambridge: Cambridge University Press.
Mellor, D. H. (1974), 'In Defence of Dispositions', *The Philosophical Review*, 83: 157–81.
Mellor, D. H. (1995), *The Facts of Causation*, London: Routledge.
Mellor, D. H. (2000a), 'The Semantics and Ontology of Dispositions', *Mind*, 109: 757–80.
Mellor, D. H. (2000b), 'Possibility, Chance and Necessity', *Australasian Journal of Philosophy*, 78: 16–27.
Mellor, D. H. (2004), 'For Facts as Causes and Effects', in J. Collins, N. Hall and L. A. Paul (eds), *Causation and Counterfactuals*, 309–23, Cambridge, MA: MIT Press.
Menn, S. (1994), 'The Origins of Aristotle's Concept of *energeia*', *Ancient Philosophy*, 14: 73–114.
Menzies, P. (1988), 'Against Causal Reductionism', *Mind*, 97: 551–74.
Menzies, P. (2009), 'Critical Notices of *Nature's Metaphysics*, by Alexander Bird', *Analysis*, 69: 769–78.
Molnar, G. (2003), *Powers: A Study in Metaphysics*, Oxford: Oxford University Press.
Mondadory, F. and A. Morton (1979), 'Modal Realism: The Poisoned Pawn', in M. J. Loux (ed.), *The Possible and the Actual*, 235–52, Ithaca: Cornell University Press.
Moravcsik, J. (1974), 'Aristotle on Adequate Explanation', *Synthese*, 28: 3–17.
Mumford, S. (1998), *Dispositions*, Oxford: Oxford University Press.
Mumford, S. (2004), *Laws in Nature*, New York: Routledge.
Mumford, S. (2005), 'Kinds, Essences, Powers', *Ratio*, 18: 420–36.
Mumford, S. (2007), 'Filled in Space', in M. Kistler and B. Gnassounou (eds), *Dispositions and Causal Powers*, 67–80, Aldershot: Ashgate.
Mumford, S. and R. L. Anjum (2011), *Getting Causes form Powers*, Oxford: Oxford University Press.
Nowak, L. (1979), *The Structure of Idealization*, Dordrecht: Springer.
Oderberg, D. (2007), *Real Essentialism*, London: Routledge.
Ott, W. (2013), *Causation and Laws of Nature in Early Modern Philosophy*, Oxford: Oxford University Press.
Pap, A. (1958), 'Disposition Concepts and Extensional Logic', in H. Feigl, M. Scriven and G. Maxwell (eds), *Minnesota Studies in the Philosophy of Science*, vol. 2, 196–224, Minneapolis: University of Minnesota Press.
Pawl, T. (2017), 'Nine Problems (and Even More Solutions) for Powers Accounts of Possibility', in J. D. Jacobs (ed.), *Causal Powers*, 106–23, Oxford: Oxford University Press.
Pietarinen, P. and V. Viljanen, eds (2009), *The World as Active Power*, Leiden and Boston: Brill.
Pietroski, P. and G. Rey (1995), 'When Other Things Aren't Equal: Saving *Ceteris Paribus Laws* from Vacuity', *British Journal for the Philosophy of Science*, 46: 81–110.
Place, U. T. (1996), 'Intentionality as the Mark of the Dispositional', *Dialectica*, 50: 91–120.
Plantinga, A. (1974), *The Nature of Necessity*, Oxford: Clarendon Press.

Plantinga, A. (1979), 'Actualism and Possible Worlds', in Loux (ed.), *The Possible and the Actual*, 253–74, Ithaca: Cornell University Press.

Plato, (1997), 'Sophist', in J. M. Cooper (ed.), *Plato. Complete Works*, 235–93, Indianapolis: Hackett.

Popper, K. R. (1957), 'The Propensity Interpretation of the Calculus of Probability, and the Quantum Theory', in S. Körner (ed.), *Observation and Interpretation*, 65–70, London: Butterworth.

Price, H. and H. Coary, eds (2007), *Causation, Physics, and the Constitution of Reality: Russell's Republic Revisited*, Oxford: Clarendon Press.

Prior, E. (1985), *Dispositions*, Aberdeen: Aberdeen University Press.

Prior, E., R. Pargetter and F. Jackson (1982), 'Three Theses about Dispositions', *American Philosophical Quarterly*, 19: 251–57.

Pruss, A. R. (2011), *Actuality, Possibility, and Worlds*, New York: Continuum.

Psillos, S. (2006), 'What Do Powers Do When They Are Not Manifested?', *Philosophy and Phenomenological Research*, 72: 137–56.

Quine, W. V. (1960), *Word and Object*, Cambridge, MA: MIT Press.

Reutlinger, A., G. Schurz, A. Hüttemann and S. Jaag (2019), '*Ceteris Paribus* Laws', in Edward N. Zalta (ed.), *The Stanford Encyclopedia of Philosophy*, Winter 2019 edn. Available online: https://plato.stanford.edu/archives/win2019/entries/ceteris-paribus/.

Robinson, H. (1982), *Matter and Sense*, Cambridge: Cambridge University Press.

Russell, B. (1976/1912), 'On the Notion of the Cause', reprinted in *Mysticism and Logic*, 173–99, London: Unwin Press.

Russell, B. (1948), *Human Knowledge: Its Scope and Limits*, London: Allen and Unwin.

Ryle, G. (1949), *The Concept of Mind*, London: Penguin.

Sanford, D. (1989), *If P, then Q: Conditionals and the Foundation of Reasoning*, London: Routledge.

Sanford, D. H. (2016), 'Determinates vs. Determinables', in Edward N. Zalta (ed.), *The Stanford Encyclopedia of Philosophy*, Winter 2016 edn. Available online: https://plato.stanford.edu/archives/win2016/entries/determinate-determinables/.

Schlick, M. (1932), 'Causality in Everyday Life and in Recent Science', *University of California Publications in Philosophy*, 15: 99–125.

Sellars, W. (1948), 'Concepts as Involving Laws and Inconceivable without Them', *Philosophy of Science*, 15 (4): 287–315.

Sellars, W. (1963a), 'Philosophy and the Scientific Image of Man', in *Science, Perception and Reality*, 1–40, London: Routledge and Kegan Paul.

Sellars, W. (1963b), 'Phenomenalism', in *Science, Perception and Reality*, 60–105, London: Routledge and Kegan Paul.

Shaffer, J. (2003), 'Is There a Fundamental Level?', *Nous*, 37: 498–517.

Shoemaker, S. (1969), 'Time without Change', *Journal of Philosophy*, 66: 363–81.

Shoemaker, S. (1980), 'Causality and Properties', in P. van Inwagen (ed.), *Time and Cause: Essays Presented to Richard Taylor*, 109–35, Dordrecht: Reidel.

Shoemaker, S. (1998), 'Causal and Metaphysical Necessity', *Pacific Philosophical Quarterly*, 79: 59–77.

Skyrms, B. (1981), 'Tractarian Nominalism', *Philosophical Studies*, 40: 199–206.
Smith, A. D. (1977), 'Dispositional Properties', *Mind*, 86: 439–45.
Smith, M. (2003), 'Rational Capacities, or: How to Distinguish Recklessness, Weakness, and Compulsion', in S. Stroud and C. Tappolet (eds), *Weakness of the Will and Practical Irrationality*, 17–38, Oxford: Oxford University Press.
Sosa, E. (1995), 'Abilities, Concepts and Externalism', in J. Heil and A. Mele (eds), *Mental Causation*, 309–24, Oxford: Clarendon Press.
Sosa, E. and M. Tooley, eds (1993), *Causation*, Oxford: Oxford University Press.
Strawson, P. F. (1959), *Individuals*, London: Routledge.
Strawson, G. (2008), 'The Identity of the Categorical and the Dispositional', *Analysis*, 68: 271–82.
Swoyer, C. (1982), 'The Nature of Natural Laws', *Australasian Journal of Philosophy*, 60: 203–23.
Taylor, H. (2018), 'Powerful Qualities and Pure Powers', *Philosophical Studies*, 175: 1423–40.
Thompson, M. (2008), *Life and Action*, Cambridge, MA: Harvard University Press.
Tooley, M. (1997), 'Causation Reductionism versus Realism', in E. Sosa and M. Tooley (ed.), *Causation*, 172–92, Oxford: Oxford University Press.
Tooley, M. (1987), *Causation—A Realist Approach*. Oxford: Oxford University Press.
Tugby, M. (2013), 'Platonic Dispositionalism', *Mind*, 122: 471–86.
Vallentyne, P. (1997), 'Intrinsic Properties Defined', *Philosophical Studies*, 88: 209–19.
Van Inwagen, P. (1975), 'The Incompatibility of Free Will and Determinism', *Philosophical Studies*, 27: 185–99.
Van Inwagen, P. (1978), 'Ability and Responsibility', *Philosophical Review*, 87: 201–24.
Vetter, B. (2013), '"Can" without Possible Worlds: Semantics for Anti-Humeans', *Philosophers' Imprint*, 13 (6): 1–27.
Vetter, B. (2014), 'Dispositions without Conditionals', *Mind*, 123: 129–56.
Vetter, B. (2015), *Potentiality: From Dispositions to Modality*, Oxford: Oxford University Press.
Vihvelin, K. (2004), 'Free Will Demystified: A Dispositional Account', *Philosophical Topics*, 32: 427–50.
Vihvelin, K. (2013), *Causes, Laws, & Free Will*, New York: Oxford University Press.
Wang, J. (2015), 'The Modal Limits of Dispositionalism', *Noûs*, 49: 454–69.
Weissman, D. (1965), *Dispositional Properties*, Carbondale: Southern Illinois University Press.
Wiggins, D. (2001), *Sameness and Substance Renewed*, Cambridge: Cambridge University Press.
Williams, D. C. (1953), 'On Elements of Being: I', *The Review of Metaphysics*, 7: 3–18.
Williams, N. E. (2010), 'Puzzling Powers: The Problem of Fit', in A. Marmodoro (ed.), *The Metaphysics of Powers: Their Grounding and Their Manifestations*, 84–105, New York: Routledge.
Williams, N. E. (2019), *The Powers Metaphysics*, New York: Oxford University Press.

Williams, N. E. and A. Borghini (2008), 'A Dispositional Theory of Possibility', *Dialectica*, 62: 21–41.

Williamson, T. (2007), *The Philosophy of Philosophy*, Oxford: Blackwell.

Wilson, J. M. (2012), 'Fundamental Determinables', *Philosophers' Imprint*, 12: 1–17.

Wilson, J. M. (2014), 'No Work for a Theory of Grounding', *Inquiry*, 57: 535–79.

Wilson, J. M. (2015), 'Hume's Dictum and Metaphysical Modality: Lewis's Combinatorialism', in B. Loewer and J. Shaffer (eds), *A Companion to David Lewis*, 138–58, Oxford: Blackwell.

Wilson, J. M. (2021), 'Determinables and Determinates', in Edward N. Zalta (ed.), *The Stanford Encyclopedia of Philosophy*, Spring 2021 edn. Available online: https://plato.stanford.edu/archives/spr2021/entries/determinate-determinables/.

Wilson, M. D. (1990), *Leibniz's Doctrine of Necessary Truth*, New York and London: Garland Publisher.

Witt, C. (2003), *Ways of Being. Potentiality and Actuality in Aristotle's Metaphysics*, Ithaca: Cornell University Press.

Yablo, S. (1992), 'Mental Causation', *The Philosophical Review*, 101: 245–80.

Yablo, S. (1999), 'Intrinsicness', *Philosophical Topics*, 26: 590–627.

Yates, D. (2015), 'Dispositionalism and the Modal Operators', *Philosophy and Phenomenological Research*, 91: 411–24.

Index

accidents 7, 54–6, 70–1, 75, 115–19, 122–3, 142, 150, 211
Achilles-heels 47, 97, 100, 101, 115–19, 122, 150, 156
actuality 14, 24, 31, 63, 184, 208, 213, 216–17
Adams, R. M. 189 n.6
Alvarez, M. 62 n.39, 172 n.17
analysis
 of abilities 2, 20, 96, 102, 111–15, 125–7, 136, 148–9
 the asymmetry of 127
 conditional analysis 91–123, 125–54
 of dispositions 36, 65 n.3, 74, 92, 94, 96–8, 102, 115–22, 160
 non-reductive 111–15, 194
 reductive 1, 74, 126 n.2, 194
Anjum, R. L. 6 n.5, 9 n.9, 37 n.10, 87 n.35, 169 n.14, 177 n.24, 203 n.27, 206 n.32
antidotes 64–5, 68, 97, 101–6, 111, 115–16, 122, 129–31, 141, 144–5, 167
Aristotle 5, 11 n.15, 14, 18, 19, 22, 24, 25, 61, 87, 162, 163, 166, 167, 176, 185, 217
Armstrong, D.M. 36 n.9, 54 n.28, 76, 87 n.34, 108 n.18, 168, 169 n.13, 200, 207, 208 n.35

Beere, J. 2 n.2
Bennett, J. 131 n.4, 132 n.5
Bird, A. 6 n.5, 9 n.9, 13 n.17, 34 n.4, 37 n.10, 56 n.31, 57 n.32, 64 n.2, 65 n.3, 71 n.12, 84 n.28, 97 n.5, 105 n.15, 145 n.16, 198 n.15, 199 nn.16–17, 200 nn.18, 21, 210 n.36, 215 nn.38–9
Blackburn, S. 205 n.31
Borghini, A. 12 n.16
Broad, C.D. 34 n.6

Butler, D. 62 n.39
butterfly-effect 13–14, 43, 71

Cambridge-change 72, 82
Campbell, K. 205 n.30
Carnap, R. 34–6, 126 n.7
Cartesian
 interaction 168
 properties of matter (see properties)
 substance 215
Cartwright, N. 34 n.3, 55 n.29, 58 n.34, 67 n.6
causalism 169
causation 1, 6 n.5, 7–11, 37 n.10, 101, 158, 168–70
Charles, D. 162 n.6, 176 n.23
Choi, S. 39 n.15, 105 nn.12, 14, 111 n.24
circularity 113–15, 216
Clarke, R. 105 n.14, 111 n.24
Coary, H. 10 n.11
concrete/abstract
 abstraction from actuality 14–15, 67, 67 n.6
 concrete as maximally specific 24
 concrete objects or substances 163, 175
 concrete situations or events 143, 145, 162, 176
 concrete worlds 187, 205
 degrees of abstractness 14, 25–6, 29, 143 n.13, 212
counterfactuals 10, 16–19, 21–2, 25, 30, 52, 63, 92–6, 97 n.6, 100, 101 n.10, 111–15, 125–35, 143–8, 155–60, 167, 178

Davidson D. 172 n.17, 173 n.19
Descartes, R. 205
determinate/determinables 67 n.5, 68–9, 76–7, 89, 103, 150, 216

determinateness and specificity   14, 21, 26, 29, 68–70, 74 n.18, 181, 211
determinism/indeterminism   134, 139 n.12, 189 n.7, 201, 203
duplication   75–80

Eagle, A.   139 n.12
Ellis, B.   6 n.5, 37 n.10, 110 n.21, 159 n.4, 180 n.28, 199 n.16, 201 nn.22–3, 210 n.36
empiricism   9, 19, 25, 34–6, 92, 160, 163, 168, 170, 186, 204
essentialism and properties   57, 199–202
essentiality and intrinsicness   75
events
    and abilities   5, 11, 14, 23, 26, 138, 156, 162–75, 185, 208, 217
    and causes   101, 172–4, 176
    as changes   22–3
    as determinables   180–1
    and dispositions   56
    fine-grained   177–80
    and the identity of modal properties   11, 16, 31, 109, 137, 150, 156–61, 171
    as intentional objects   15, 60
    and observations   35, 92–3, 99, 114, 141, 155, 156, 159
    as stimuli   103, 108, 114
    their role in ontology   186, 212–13
explanation
    causal   7, 10, 11 n.15
    of laws   53, 116, 119
    metaphysical   3, 27–9
    by redescription/identification of modal properties   10 n.13, 11 n.14, 159

Fara, M.   34 n.5, 45 n.22, 99 n.9, 105 n.14, 144 n.15
Fine, K.   126 n.3
finks and reverse finks   98–100, 104–11, 115, 122, 136, 160
frequency   41, 59, 122–3
Funkhouser, E.   68 n.8

Gendler-Szabó Z.   133 n.7
generality and abilities   14, 52, 63, 67, 73–4, 91, 102–6, 125, 128–9, 141, 143–8, 209

Glynn, L.   9 n.9, 139 n.12
Goldman, A.   98
Goodman, N.   35–6, 37 n.11
Griffin, M. V.   189 n.7

habits/habitual behavior   11–14, 34, 41, 49, 59–61, 92, 123, 153
Harré, R.   6 n.5, 33 n.1, 37 n.10, 85, 165, 169 n.14
Hawthorne, J.   200 n.19
Hegel, G.W.F.   9 n.10, 26, 33, 60 n.36, 67, 146, 155, 183, 184, 216
Heil, J.   79 n.25, 87 nn.34–5, 143 n.14, 202 n.25, 206 n.33, 215 n.39
Hempel, C.   146 n.18
Hintikka, J.   2 n.2
Hoefer, C.   137 n.10, 139 n.12
Holton, R.   215 n.39
Horgan, T.   28 n.28
Humberstone, I. L.   75 n.20, 77 n.23
Hume, D.   4, 7, 9, 132, 168, 174, 200, 204, 212
Huoranszki, F.   6 n.5, 54 n.28, 61 n.37, 70 n.11, 84 nn.29–30, 111 n.22, 115 n.26, 152 n.22, 153 n.23, 189 n.7, 196 n.12, 198 n.14, 202 n.25
Hüttemann, A.   90 n.38
Hyman, J.   172 n.17

idealism   212–14
Ingthorsson, R.   215 n.38
inhibition   141, 144–5, 148, 160, 165, 177
intentional
    action/behavior   45, 60–2, 122, 172 n.18, 173
    states and objects   15–16, 60, 155, 158 n.3, 163
interaction   7–10, 12, 22, 24, 57, 84–5, 87–9, 143–9, 151, 166 n.8, 167, 175–9, 196, 205, 207, 209–13
irreducibility of the modal properties   72, 108, 115

Jacobs, J. D.   4 n.3, 12 n.16
Johnston, M.   97 n.5, 99 n.8, 156 n.1
Jubien, M.   126 n.2

Kant, I.   9 n.10, 183, 189 n.8
key and lock example   81–3

Kim, J.   22 n.22, 75 n.21
know-hows and skills   16, 53, 153–4
Koslicki, K.   29 n.30
Kosman, L. A.   162 n.6

Langton, R.   74 n.16, 75 nn.20–1
laws of nature   1, 2, 4, 6–7, 9–12, 30, 34, 37, 53–8, 69–70, 76, 89 n.37, 111 n.22, 117–19, 198 n.14, 201 n.23, 202–3, 208, 212, 216
Leibniz, G. W.   3, 24 n.24, 162 n.6, 189
Leibniz's law   207
Levey, S.   24 n.24
Lewis, D.   1, 1 n.1, 18 n.20, 38, 39 n.16, 65 n.3, 69 n.10, 74, 74 nn.16–17, 75 nn.20–1, 87 n.34, 105, 105 n.13, 106–8, 108 n.17, 109, 109 n.19, 110, 110 n.20, 111 nn.23–4, 111–15, 115 n.27, 132, 132 n.5, 134, 143 n.14, 169 n.13, 171–3, 174 n.20, 186, 187 nn.4–5, 187–8, 197 n.13, 198–200, 208 n.35, 216 n.41
Lierse, C.   110 n.21, 159 n.4, 180 n.28
Locke, J.   23 n.23, 67 n.6, 162 n.6, 197
Loewer, B.   111 n.22
Lombard, L.   22 n.22, 161 n.5
Lowe, E. J.   15 n.18, 152 n.22, 180 n.27, 215 n.39

Mcdonald, C.   22 n.22
Mackie, J. L.   23 n.23, 97 n.6, 108 n.18, 158–60, 169 nn.13–16, 208 n.35, 217
McKitrick, J.   69 n.10, 72 n.13, 81 n.27, 110 n.21, 111 n.24, 168 n.12, 180 n.27
Madden, E.H.   6 n.5, 37 n.10, 85
manifestation
    of abilities   83, 122, 136, 155, 165, 181
    as causal response   101, 108–9, 168–9, 171–2, 174
    as contribution   177–9
    of dispositions   55–62, 87, 116, 150
    as event   94, 103, 156–60, 180
    as a feature of an object   150
    as non-change   177
    the original concept of   25, 92–3, 113–14, 141, 168, 170
    as qualitative change   161–3

Manley, D.   47 n.24, 97 n.7, 157 n.2
Marmodoro, A.   8 n.6, 87 n.35
Marshall, D.   75 n.20
Martin, C. B.   15 n.18, 87 nn.34–5, 98, 107, 108 n.17, 111–12, 122, 143 n.14, 158 n.3, 169 n.14, 176, 206 n.33
masks   20, 22, 97, 99, 101–7, 111, 115, 117–18, 122, 129–30, 141, 144–5, 160
materialist metaphysics   168
Mates, B.   189 n.6
matter   31, 42 n.19, 73, 79, 83, 86, 111, 163, 176 n.22, 185, 204–13, 217–18
Mellor, D.H.   33, 33 n.2, 87 n.33, 93 n.1, 108 n.17, 110 n.21, 112, 139 n.12, 151–2 n.21, 169 n.16, 180 n.28, 194, 194 n.11, 216 n.41
Menzies, P.   57 n.32, 115 n.27, 200 n.20
mimics   22–3, 98–9, 101, 151, 155–7, 159–60, 162–3, 171
Molnar, G.   15 n.18, 81 n.27, 85 n.31, 108 n.17, 110 n.21, 149, 169 n.14, 177 n.24, 180 n.27, 210 n.36
Mondadory, F.   4 n.3
Moravcsik, J.   11 n.15
Morton, A.   4 n.3
multitrack dispositions   180–1
Mumford, S.   6 n.5, 9 n.9, 37 n.10, 57 n.32, 87 n.35, 93 n.1, 108 n.17, 169 n.14, 177 n.24, 203 n.27, 206 n.32

natural
    ends   59, 121
    kinds   7, 54–8, 60, 118, 120–2, 159 n.4, 180 n.28, 199, 211
    laws (*see* laws of nature)
    necessity   6, 37
    properties   69–70, 83–4, 143, 198, 211
necessity
    and essentialism   199, 201
    hypothetical   202
    in nature   1, 2 n.2
    necessary connection   1, 2, 7, 33, 37 n.10, 132, 192, 203
    necessary truth   3, 55, 189 n.6
    a theory of   188

and 'would-conditionals'   132
necessitarian world   3, 189, 201–2
Nowak, L.   67 n.6

Ott, W.   15 n.18

pandisposionalism   204, 206
particulars   59, 76, 85–6, 89–90, 118 n.28, 121, 176 n.22, 179, 184, 186, 192, 200, 204, 207, 212, 218 n.44
Pawl, T.   12 n.16
Pfeifer, K.   15 n.18, 158 n.3
phenomenalism   35–7, 92–5, 114, 163, 167, 168, 170–1, 213
Pietroski, P.   119 n.29
Place, U. T.   15 n.18, 87 n.34
Plato   8, 8 n.7, 61, 72, 86
Popper, K. R.   72 n.13
possibility
    contingent   2, 4, 12–14, 26, 29–31, 34, 38, 46 n.23, 64, 67, 70, 92, 95, 103, 113, 119, 126, 135, 147, 184–5, 191, 193, 213, 217
    and contradiction   3, 130, 134–5, 137, 189–92
    and worlds   2–4, 14, 17–19, 24, 31, 69, 79, 91, 95, 126, 132, 183–98, 200–1, 203
    powers   1, 2, 5–10, 12, 15 n.18, 23 n.23, 33, 34 n.6, 37, 39 n.13, 57 n.32, 72, 79 n.25, 85, 87, 89 n.36, 98, 108 n.17, 111, 115, 149, 151, 165, 169, 199–203
Price, H.   10 n.11
priority
    of the actual   185, 217–18
    explanatory   18
    metaphysical   27–9
probability   72 n.13, 99, 134, 137–40, 151
properties
    abundant or sparse   69–70
    Cartesian   210–11
    categorical   4–6, 23–4, 33, 53, 94, 111, 168–70, 185, 191, 206–9, 212, 214, 217–18
    first-order modal   2, 4, 8, 13, 17, 53–4 n.28, 72, 74, 91, 94–5, 100, 108, 126, 183–5, 188, 207, 214–15, 217–18
    first-order nonmodal   115, 197, 200

second-order properties   4, 8, 115, 196–7, 200
statistical   6 n.5, 37, 41, 48–9, 90
Psillos, S.   8 n.6

qualitative
    changes   31, 155, 159, 161–7, 171, 173, 176, 178, 180
    non-changes   164, 166
    properties   185, 191, 203–4, 206, 212 n.37, 213, 214, 216 n.42, 217–18
    specificity   63–4
Quine, W.V.   18 n.19, 50 n.27

rationalism and modal properties   217
realism about properties   4, 7, 17, 20, 72, 74, 108 n.17, 215
reasoning and the ascription of modal properties   13, 21, 72, 100, 125, 129–30, 140–8, 150
recombination of properties   23, 192–8
reduction   17, 20, 92, 95, 112, 126, 178, 194
regress of pure powers   214
Reutlinger, A.   56 n.30
Rey, G.   119 n.29
Robinson, H.   212 n.37, 215 n.38
Russell, B.   10 n.12, 168 n.11, 204
Ryle, G.   5, 5 n.4, 10, 10 n.13, 18 n.19, 34 n.5, 36, 41, 41 n.17, 53, 53 n.28, 61, 159, 168, 170, 180 n.29, 181

Sanford, D.   68 n.8
Schlick, M.   168 n.11
sensitivity and degrees of extrinsicness   74, 80–5
Shaffer, J.   84 n.30
Shoemaker, S.   33 n.1, 72 n.14, 81 n.27, 166 n.8, 198 n.15, 199 n.17, 201 n.23
Smith, A.D.   72 n.13, 98, 99 n.8, 111 n.24, 156 n.1
Smith, M.   34 n.5
Socrates   72, 86
Sosa, E.   66 n.4
Strawson, G.   206 n.33
Strawson, P.F.   27 n.27
substance
    as agent or persisting particular   172, 190, 204

chemical   9, 16, 54, 56–7, 64–6, 72,
    74, 77, 88, 116, 119, 144, 164 n.7,
    170, 175, 176 n.22, 205, 213, 217
Swoyer, C.   199 n.17
systems
    as bearers of intrinsic abilities   85–90
    as ensembles of objects or particles   4,
        29, 40, 67, 85–6, 89–90, 121–2, 186,
        207, 213
    physical   7, 58, 75–6, 115 n.26, 122,
        174, 202, 205, 217
    of properties   144, 215–16

Taylor, H.   206 n.34
teleological dispositions   12, 59–61, 92,
    121–2, 150
tendency and dispositions   2, 6 n.5,
    13–14, 26, 34, 41–3, 46–9, 52, 56–9,
    61, 66, 92, 117, 119, 120, 122–3,
    139–40, 144, 153, 211
Thompson, M.   50 n.26
Tooley, M.   89 n.37

Vallentyne, P.   75 n.20
Van Inwagen, P.   173 n.19, 202 n.24
verificationism   19, 35–7, 92, 126
Vetter, B.   12 n.16, 34 n.5, 37 n.11,
    38 n.12, 39 n.13, 49 n.25, 50 n.27,
    63 n.1, 72 n.13, 81 nn.26–7
Vihvelin, K.   34 n.5

Wasserman, R.   47 n.24, 97 n.7, 157 n.2
Weatherson, B.   75 n.20
Wiggins, D.   12 n.16
Williams, D. C.   205 n.30
Williams, N.E.   12 n.16, 89 n.36, 206 n.33
Wilson, J. M.   29 n.29, 68 n.8, 181 n.30,
    192 n.9
Wilson, M. D.   189 n.8
Witt, C.   19 n.21
Wittgenstein, L.   168

Xantippe   72

Yablo, S.   75 n.20, 181 n.30

www.ingramcontent.com/pod-product-compliance
Lightning Source LLC
Chambersburg PA
CBHW062144300426
44115CB00012BA/2034